T0338270

BANDWIDTH EFFICIENT CODING

BANDWIDTH EFFICIENT CODING

JOHN B. ANDERSON
Ericsson Professor in Digital Communication
Electrical and Information Technology Department
Lund University
Lund, Sweden

IEEE SERIES ON
DIGITAL
& MOBILE
COMMUNICATION

John B. Anderson, *Series Editor*

IEEE PRESS

WILEY

Published by John Wiley & Sons, Inc., Hoboken, New Jersey
Published simultaneously in Canada

For general information on our other products and services or for technical support, please contact our Customer Care Department within the United States at (800) 762-2974, outside the United States at (317) 572-3993 or fax (317) 572-4002.

Wiley also publishes its books in a variety of electronic formats. Some content that appears in print may not be available in electronic formats. For more information about Wiley products, visit our web site at www.wiley.com.

Library of Congress Cataloging-in-Publication Data is available.

ISBN: 978-1-119-34533-6

Printed in the United States of America

10 9 8 7 6 5 4 3 2 1

CONTENTS

Preface   ix

1 Introduction   1

   1.1 Electrical Communication, 2
   1.2 Modulation, 4
   1.3 Time and Bandwidth, 9
   1.4 Coding Versus Modulation, 13
   1.5 A Tour of the Book, 14
   1.6 Conclusions, 15
   References, 16

2 Communication Theory Foundation   17

   2.1 Signal Space, 18
   2.2 Optimal Detection, 24
        2.2.1 Orthogonal Modulator Detection, 24
        2.2.2 Trellis Detection, 29
   2.3 Pulse Aliasing, 35
   2.4 Signal Phases and Channel Models, 37
   2.5 Error Events, 43
        2.5.1 Error Events and d_{min}, 43
        2.5.2 Error Fourier Spectra, 48
   2.6 Conclusions, 50
   Appendix 2A: Calculating Minimum Distance, 50
   References, 56

3   Gaussian Channel Capacity   **58**

3.1 Classical Channel Capacity, 59
3.2 Capacity for an Error Rate and Spectrum, 64
3.3 Linear Modulation Capacity, 68
3.4 Conclusions, 72
Appendix 3A: Calculating Shannon Limits, 73
References, 77

4   Faster than Nyquist Signaling   **79**

4.1 Classical FTN, 80
    4.1.1   Origins of FTN, 80
    4.1.2   Definition of FTN Signaling, 81
    4.1.3   Discrete-Time Models, 86
4.2 Reduced ISI-BCJR Algorithms, 87
    4.2.1   Reduced Trellis Methods: The Tail Offset BCJR, 89
    4.2.2   Reduced-Search Methods: The M-BCJR, 93
    4.2.3   The ISI Characteristic, 99
4.3 Good Convolutional Codes, 101
    4.3.1   Binary CC Slope Analysis, 102
    4.3.2   Good Binary Modulation Codes, 105
    4.3.3   Good Convolutional Codes for 4-ary Modulation, 107
4.4 Iterative Decoding Results, 110
4.5 Conclusions, 114
Appendix 4A: Super Minimum-Phase FTN Models, 115
Appendix 4B: Good Convolutional Codes for FTN Signaling, 116
References, 124

5   Multicarrier FTN   **127**

5.1 Classical Multicarrier FTN, 128
5.2 Distances, 134
    5.2.1   Finding Distances, 134
    5.2.2   Minimum Distances and the Mazo Limit, 136
5.3 Alternative Methods and Implementations, 138
5.4 Conclusions, 143
References, 143

6   Coded Modulation Performance   **145**

6.1 Set-Partition Coding, 146
    6.1.1   Set-Partition Basics, 146
    6.1.2   Shannon Limit and Coding Performance, 150
6.2 Continuous Phase Modulation, 153
    6.2.1   CPM Basics, 153
    6.2.2   Bits per Hz-s and the Shannon Limit in CPM, 157
    6.2.3   Error Performance of CPM, 158

6.3 Conclusions for Coded Modulation; Highlights, 161
References, 161

7   Optimal Modulation Pulses   163

7.1 Slepian's Problem, 164
    7.1.1   PSWF Pulse Solution, 165
    7.1.2   Gauss and Gauss-Like Pulses, 169
    7.1.3   Occupancy of Linear Modulation with FTN, 172
    7.1.4   PSWF and Gauss Linear Modulation with FTN, 175
7.2 Said's Optimum Distance Pulses, 177
    7.2.1   Linear Programming Solution, 178
    7.2.2   Optimal Modulation Tap Sets, 180
    7.2.3   Coded and Uncoded Error Performance, 182
7.3 Conclusions, 185
Appendix 7A: Calculating the PSWF, 185
Appendix 7B: Optimum Distance Tap Sets, 188
References, 188

Index   190

PREFACE

Coded communication that is both bandwidth and energy-efficient has traversed a long and twisted path. As with much else, it began in Shannon's first channel coding paper: That work allows for any number of databits to be crammed into a unit of bandwidth, and some important present-day formulas are there. Still, coding in its first 30 years was viewed as a wideband technique, that made possible reduced energy in return for extra bandwidth. In this way, we have reached the planets and beyond.

On Earth today we need to save bandwidth. Coding that did not increase bandwidth began to appear in the 1970s under the heading of coded modulation. The realization that such was possible was a shift in our thinking. But there were puzzles. Coding led to large energy savings in wideband transmission, but did the same savings exist at narrowband? Intersymbol interference (ISI) problems attracted interest, and methods for removing ISI more and more resembled decoding algorithms, but was ISI a form of coding, and if it was not, how did it differ? How to think about bandwidth was an ongoing puzzle and common ways of measuring it sometimes led to signaling that seemed to violate Shannon's capacity. What role orthogonal transmission pulses play was another puzzle; many thought that orthogonal transmission was sufficient to describe all interesting methods. All of this became more mysterious as signaling bandwidth narrowed.

This book is the culmination of a long effort by the author and others to allay these mysteries. To do so requires broadened and more careful ideas of modulation, coding, and bandwidth. Without them the book cannot succeed, and much attention has been given to these concepts.

There have been controversies along the way. Mazo's idea of faster than Nyquist signaling was not easily accepted. How could the Nyquist criterion be violated? The way out of the paradox lay in more careful definitions. In more recent years, some have objected to ideas that appeared to violate capacity. This too was resolved by more careful definition. Let it be said at the outset that nothing in this book violates either Shannon or Nyquist. Their ideas are safe and strong as ever.

Overall, the book seeks to resolve mysteries, then measure the performance of real methods, then make it possible for others to repeat the tests and adapt the methods to their own use. With regard to the last, the book has several special features. An effort is made to list concrete, good codes. Matlab routines are given that measure minimum distance, probability of error, and Shannon limits. ISI configurations are given that achieve set bandwidths. Particular attention is given to receiver algorithms. These are generally simple in principle, but there are many troublesome details in their implementation. This is particularly true of algorithms that handle ISI, and readers will not get the performances in the book unless most of these details are attended to. It is often true in narrowband coding that the Devil is in the Details. As a rule, one can ignore many if the transmission energy is raised, but coding is after all about saving energy, and ultimate performance is thus a matter of details.

In tracing the story of bandwidth efficient coding, I have attempted to show its historical development as well as explain its ideas. However, I am not a historian and I apologize to the many contributors that are not referenced by name and may have been overlooked. The priority in citations is first to establish the advent of ideas, second to give the reader a good source of further knowledge, and only third to show the breadth of contributors.

In this and other work, I have had the benefit of a great many research colleagues over the years. Space limits me here to acknowledge only those who made special contributions to the book and to the ideas of bandwidth efficient coding. I would like to mention first some early pioneers. James Mazo of Bell Laboratories was a fundamental contributor who 40 years ago devised the attitude behind and the provocative name Faster than Nyquist Signaling. B, C, J, and R (Lalit Bahl, John Cocke, Frederik Jelinek, Josef Raviv) performed early iterative decoding work and devised the eponymous BCJR algorithm, without which decoding cannot function. I particularly acknowledge Frederik Jelinek, who was my Ph.D. supervisor at Cornell University; despite rumors to the contrary, he and I worked on a completely different topic, but the BCJR and iterative decoding were near at hand. Prof. John Proakis of Northeastern University, Boston has contributed in diverse ways to intersymbol interference problems over the entire history of the subject and is an inventor of the optimal detector for it. Prof. Lars Zetterberg of Royal Institute, Stockholm, believed in bandwidth efficient coding from the 1960s and inspired many on the Swedish scene to pursue these problems. Catherine Douillard, Claude Berrou and coworkers proposed the turbo equalizer, which is still the basis of most really bandwidth efficient coding. Last I acknowledge David Slepian, who was a friend and inspiration to many of us in the years 1960–1985, and who more than anyone else straightened out the puzzle of signal bandwidth. Both David (2007) and Frederik Jelinek (2010) have passed away and cannot know what their work became.

In more modern times, I was fortunate to work with a number of talented research students at Rensselaer Polytechnic Institute and Lund University whose work figures in this book. These include Fredrik Rusek, who derived detailed capacity results and much else, Adnan Prlja, who carried out early iterative receiver tests, and Mehdi Zeinali, who searched for and found good convolutional codes, sometimes after hours and without a salary. During the 1980s and 1990s, major contributions were made by Nambirajan Seshadri and Krishna Balachandran, who attacked the intersymbol interference receiver issues that were then outstanding. Amir Said in 1994 finally placed energy and bandwidth of intersymbol interference on a firm theoretical footing.

In recent years, I have had many discussions and argumentations with European colleagues. Particularly, I would like to acknowledge Göran Lindell and Micha Lentmaier at Lund, Nghia Pham of EUROSAT Paris, and Joachim Hagenauer at the Technical University of Munich. Special thanks are due to my editorial colleagues at Wiley-IEEE Press over the years and to my editor for this book Mary Hatcher.

It is a mistake to think that ideas come from a single time and place, and to the many not cited here and whose influence was more indirect, I say a grateful thank you.

It is a pleasure to recognize research support that played a role in the work presented here. The Swedish National Research Agency (Vetenskapsrådet) supported the work as part of annual grants over the years 2000–2013. The Swedish Foundation for Strategic Research (Stiftelse för Strategisk Forskning) was instrumental with its long-term funding of the Lund Center for High-Speed Wireless Communication starting in 2006. I would especially like to acknowledge the L.M. Ericsson company, which established the Chair in Digital Communication that I occupy at Lund University. These three organizations have been mainstays of basic engineering research in Sweden, to the great benefit of that country, and indeed the world.

Lund, Sweden   JOHN B. ANDERSON

1

INTRODUCTION

PROLOGUE

How does one transmit information when bandwidth is expensive? One can explore new wavelengths. One can use multiple antennas to distinguish more pathways. One can design better modulation and coding for each pathway. This book is about the last.

According to the laws of nature, sending bits of information via an electrical medium requires two main resources: energy and bandwidth. Each bit needs energy, and just as in soccer football, a certain energy is needed to hit the ball far enough. Adding bandwidth to football needs new rules. Suppose every player has a ball and all must pass through a narrow passage on the way to the goal. How difficult is it to reach the goal?

Electrical communication is a game too, played according to nature's laws. In the first half of the twentieth century, several of these became clear. First, much less energy is needed per bit if the signal bandwidth is widened. Second, a certain type of energy pulse, called "orthogonal," is easier to process. These facts were well established in 1949 when Claude Shannon published something different, a formula for the ultimate capacity of a set bandwidth and energy to carry information. The key to approaching that limit is coding, that is, imposing clever patterns on the signals. As much as 90% of signal energy can be saved compared to rudimentary methods, or the information can be sent much further. In principle, whatever configuration of energy and bandwidth was available, it could be coded.

Bandwidth Efficient Coding, First Edition. John B. Anderson.
© 2017 by The Institute of Electrical and Electronics Engineers, Inc. Published 2017 by John Wiley & Sons, Inc.

Nonetheless, those who designed codes in the following decades thought of them, consciously or not, as trading more bandwidth for less energy. Even with a simple trade, a signal could travel much further if its bandwidth were scaled up. In this way—and only this way—signals reached Mars and even Pluto. Neither was crowded with players, so that wide bandwidth signals were practical. There will soon be rovers from several nations on Mars, so that allotting wide bandwidth to each is not quite so practical. Here on Earth the situation is more desperate. Such services as cellular wireless and digital video must share a very crowded spectrum; furthermore, governments have discovered they can force providers to pay astonishing prices for bandwidth. Today, bandwidth costs far more than energy. Everyone needs to minimize it.

In the 1970s, methods of signal coding were discovered that did not increase bandwidth; signals could travel farther and easier without scaling up bandwidth. This was clear in Shannon's 1949 formula from the first day, but the concepts took time to sink in. What his formula really says is that coding leads to large savings no matter what the combination of energy and bandwidth. Today, with bandwidth costly, we want to work at narrow bandwidth and higher energy, not the reverse. The problem is: We know little about how to design the coding. The purpose of this book is to take that next step. How should efficient signal coding work when very narrow bandwidths per bit are available?

1.1   ELECTRICAL COMMUNICATION

This chapter introduces important and potentially troublesome concepts in a mostly philosophical way. Among them are bandwidth and time, pulse shapes, modulation, and coding. The needed formal communication background is Chapter 2, and Shannon's theory appears in Chapter 3. Certain concepts need some evolution to fit modern needs. The outcome of the book is that practical coding schemes exist that work well in a very narrowband world.

Communication transmits messages through time and space. In this book, the medium is electrical signals, and we will restrict ourselves to radio. For our purposes, messages are composed of symbols, and we want to transmit these accurately and efficiently through the physical world. In communication engineering, sending a symbol has three basic costs, energy, bandwidth, and implementation complexity. Each trades off against the others. Once the three costs are set the measure of good performance is most often probability of symbol error.

Even a century ago, it was clear enough that error may be reduced by spending more energy. A basic fact of communication, that first became evident with FM broadcasting, is that for the same performance, energy may be traded for bandwidth. Error-correcting codes, as first used, were thought of as a manifestation of this fact: By transmitting extra check bits, energy could be saved overall. But Claude Shannon's 1949 work [7] implied that energy or bandwidth or both could be reduced while maintaining the same error performance. Each combination of energy and bandwidth had a certain capacity to carry symbols but alas, there was no free lunch, and rapidly diverging "coding" complexity is required to approach this capacity. Today, energy, bandwidth, and complexity are a three-way trade-off.

Since 1949 there have been 60 years of progress and a rich research literature, but coding and its attendant complexity have been associated mainly with relatively low energy and wide bandwidth on a per data bit basis. Economic necessity and some changes in physical transmission paths are forcing changes in that thinking. Two examples are wireless networks and satellite digital broadcasting. Successive mobile network generations have offered more bits per second to their customers and the upcoming fifth Generation Systems will employ a variety of methods to offer even more. Among them are better antennas and shorter paths, both of which increase the symbol energy, making possible narrower bandwidth. According to theory, there is just as much to gain from coding and complexity in this narrowband/high-energy regime as we have enjoyed in wider band applications.

We move now to fundamental concepts and a sketch of the book. Knowledge of history helps make sense of a subject, and so we offer some high points, together with some important published works. The prerequisites for the book are first courses in probability, Shannon theory, signal space theory, and coding methods of the traditional type. Our own treatment of these subjects will be to extend their results to the new narrowband field.

Bandwidth and Coding. Coding, bandwidth, and their interaction lie at the heart of this book.

Coding theory is built upon abstract models, nonphysical concepts such as channel models, information, symbols, and arithmetic operations. Although it is not necessary to its mathematics, the theory can suggest physical conclusions by starting with signals and symbols that are avatars to physical channels and transmissions. In this book, the physical channel is always the white noise linear channel, wherein a white noise stochastic process with a certain power spectrum $N_0/2$ watts/Hz (the *noise*) adds to a analog function of time (the *signal*). Something is needed to convert abstract symbols to the analog domain; this is a *modulator*. This last is explored in Section 1.2. A modulator–demodulator by itself has a certain probability of error. The import of Shannon's work is that another two boxes, the *encoder–decoder*, can be added that can in principle reduce the error rate virtually to zero, provided that the energy–bandwidth combination is sufficient. The distinction between coding and modulation is a tricky one, taken up initially in Section 1.4, and the Shannon theory tools we need are in Chapter 3. That chapter starts from the core result in the 1949 paper, which is the capacity of the additive white Gaussian noise (AWGN) channel. In this channel, an independent Gaussian-distributed noise value with mean zero and variance $N_0/2$ is added to a real signal value. Initially, there were no analog signals; part of Shannon's genius was making the critical jump from the abstract AWGN channel to the channel with analog signals and bandwidth that we need in this book.

While it is true that the promise of narrow band/high-energy coding was already clear in 1949, we have spent most of the time since then developing techniques in the opposite regime. During its first 25 years, coded communication was primarily about parity-check codes and the binary symmetric channel. Codewords contained extra "redundant" bits, there being no other way to distinguish codeword bits from customer bits. With such codes nearly universal, one could easily fall into the belief that coding mostly exchanged extra bits (i.e., bandwidth) for coding gain (reduced energy). With the advent of coded modulation in the mid 1970s, cracks appeared in this belief. New transmission methods such as continuous-phase modulation (CPM) and set-partition coding appeared, which reduced energy without bandwidth expansion, and sometimes did this without redundant bits. By the late 1970s, it was clear that codes could even be used to *reduce* bandwidth, or even bandwidth and energy both.[1] These new methods eventually broke through the coding equals bandwidth-expansion belief, and doubled or tripled the rate of coded communication in a given bandwidth. Today, we would like to double or triple it once more.

More changed in the 1970s and 1980s than the bandwidth/energy regime. If coding was not redundant bits, then what was it? If CPM, which seemed to be modulation, and convolutional codes both required a trellis decoder of similar size, then were they not both coding schemes? Set-partition codes contained a convolutional code within them; was that the code, or was something else? Should removal of intersymbol interference (ISI) be called decoding? It became clear that coding needed more careful definition if there were to be reliable structure and language for our work.

So to bandwidth itself. To the philosophical, bandwidth is a measure of *change-ability*, referenced to a unit of time. In the popular mind, it is the number of data bits a system such as a telephone handles, again per unit time. To a prosaic communications engineer, it is simply the width of a Fourier transform. These views are not really different: A signal can only convey information by changing, the more change the more information, and a signal with a narrow Fourier spectrum is one that changes slowly. All three views carry within them a unit of time, in MKS units the second. These views reflect our life experience, but communication theory is based on the *product* of changeability and time, the total change however accumulated. To send a certain quantity of information requires a certain accumulation. The product unit is the Hertz-second, a subtle concept explained more fully in Section 1.3.

This book is about more efficient transmission via coding when the relative bandwidth is narrow. The channel is linear and nonchanging, with simple white noise. We can look forward to second-generation research in synchronization, fading and nonlinear channels in the future, but they are not here yet.

1.2   MODULATION

Modulation is the conversion of symbols to physical signals. In this book, they are analog signals. Effectively, it is digital-to-analog conversion. We may as well assume that data to be transmitted arrive as bits, and the modulator then accepts these $\log_2 M$ bits at a time. We normally think of a modulator as accepting $\log_2 M$-bit groups and

---

[1] The author recalls adding some provocation to conference presentations in the 1970s by suggesting that the coding schemes presented did not increase bandwidth. Fortunately, the result was only laughter.

applying some process to each one. Most often the modulation process works from a set of M alternatives; for example, it may produce M tones or phases or amplitudes. We associate each M-fold alternative with a piece of transmission time, the *symbol time*, denoted T_s, or when no confusion can result, simply T. The average transmitted energy during T_s is the *symbol energy E_s*.

Referring spectra and energy back to the data bits divides their values by $\log_2 M$. This leads to the soundest analytical picture, but the M that best exploits the channel resources is often nonbinary. We will reserve *transmission symbol* to mean this M-ary set of alternatives.

Within this framework, a formal definition of modulator for this book is "*A conversion of all possible M-ary symbol sequences to analog signals by repeated application of a fundamental operation to each symbol*".

Most modulators in use today are *pulse* modulators, meaning that they associate each symbol with a pulse according to

$$s(t) = \sqrt{E_s} \sum_n u_n h(t - nT), \tag{1.1}$$

where u_n is the symbol and $h(t)$ is the *base pulse*. The symbols are pulse amplitudes and by convention they are independent random variables with zero mean and unit variance; consequently, E_s in Eq. (1.1) is the average symbol energy. The process here is called *linear modulation*, because the pulses simply add. In the classic modulator, the pulses are T-orthogonal, meaning

$$\int h(t)h(t - nT)\, dt = 0, \qquad n \text{ an integer}, n \neq 0. \tag{1.2}$$

The great majority of modulators in applications heretofore are linear and use orthogonal pulses. Nonlinear modulators have some use, and classic examples are frequency-shift keying and the CPM signaling in Chapter 6. There is little loss in Shannon theory from the linearity requirement, but the same is not true with orthogonality. There is evidence that it leads to loss in many situations and most new methods in this book dispense with it. Its advantage is that it leads to a simple optimum detector; this is explored in Section 2.2.

To maintain the unit variance property on u_n, a binary symbol alphabet needs to be $\{+1, -1\}$. A uniformly spaced 4-ary modulator has alphabet $(1/\sqrt{5})\{\pm 3, \pm 1\}$ and an 8-ary $(1/\sqrt{21})\{\pm 7, \pm 5, \pm 3, \pm 1\}$. These three standard modulators will be referred to by their traditional names 2PAM, 4PAM, and 8PAM, where PAM means pulse-amplitude modulation. There can be a small advantage from nonuniform spacing, especially in bandpass schemes, but we will not pursue this in the book.

Equation (1.1), without a sinusoidal carrier, is said to be in *baseband* form. The base pulse $h(t)$ and the signal $s(t)$ ordinarily have a lowpass spectrum. Most applications employ carrier modulation, which is the same except that the spectrum of $s(t)$ is translated up by the f_c Hz, the carrier frequency. The translation is performed through multiplication by either $\sin 2\pi f_c t$ or $\cos 2\pi f_c t$, and the final signal has the form

$$s(t) = \sqrt{2E_s}\,[I(t)\cos 2\pi f_c t - Q(t)\sin 2\pi f_c t]. \tag{1.3}$$

Here $I(t)$ and $Q(t)$ are both baseband, that is, lowpass, signals called respectively, the in-phase and quadrature signals, and the outcome is a signal with a narrow spectrum centered at f_c Hz. $I(t)$ and $Q(t)$ satisfy

$$\mathcal{E}\Big[\int_T [I^2(t) + Q^2(t)]\,dt\Big] = 1,$$

where $\int_T$ is over a signal interval; consequently E_s is the symbol energy, this time for a dual symbol and two signals. Equation (1.3) is said to be a *passband* signal, written in the in-phase and quadrature, or I/Q, form.[2] Observe that one passband signal corresponds to two baseband signals. Each can take an independent Eq. (1.1) and they can be detected independently if $\cos 2\pi f_c t$ is known. Any passband signal can be constructed in this way.

Passband signals are essential because they allow many signals to be sent through the same medium and the properties of different wavelengths to be exploited. But for several reasons we will treat primarily baseband signals in this book. The chief one is that with perfect phase synchronization, $I(t)$ and $Q(t)$ are obtainable and independent and there is no reason to add the complexity of the passband form. A passband notation is a statement that there is imperfect synchronization or that there exist channel distortions that affect I and Q differently. These are interesting topics, but the schemes in this book have not yet reached that state of the art.

Pulse Properties. At first it may seem that a practical pulse is one that takes place wholly in its own interval, but it became clear by the 1950s that not much reduction in implementation comes from this and the bandwidth properties are far worse. The serious study of pulse shapes began with Harry Nyquist in 1924 [5], who studied pulses that have zeros at all integer multiples of T. This property is now called the *Nyquist pulse criterion* (NPC). He proved that a sufficient condition for a symmetric NPC pulse is

Property 1.1 (**The Spectral Antisymmetry Condition**) *A sufficient condition that a symmetrical $h(t)$ be NPC is that $H(f)$, the Fourier transform, is antisymmetric about the points $[H(f), f] = [H(0)/2, 1/2T]$ and $[H(0)/2, -1/2T]$.*

Note that a symmetric pulse has a real, symmetric transform. Later researchers found the necessary and sufficient condition and removed the requirement that $h(t)$ be time symmetric. These issues are discussed in introductory texts ([1], Section 2.2; [2], [3]). However, there is little reason in theory or in applications to abandon

---

[2]A common alternate framework takes $s(t)$ as the real part of $h(t)\exp j2\pi ft$, where $h(t)$ is now complex. See, for example, the text [3]. The method is used because complex numbers mimic the needed operations, not because there are unreal signals.

symmetric pulses, and we will use them and the antisymmetry condition almost exclusively.

In a later paper [6], Nyquist observed that there seemed to be a lower limit of about $1/2T$ Hz to the bandwidth of an NPC pulse. Research by others eventually developed the following theorem and the closely allied sampling theorem:

Property 1.2 *The bandwidth of any NPC pulse with zeros at nT cannot be narrower than $1/2T$ Hz, and the narrowest pulse is $h(t) = A \operatorname{sinc}(t/T)$, A is a real constant.*

Here $\operatorname{sinc}(x)$ is defined as $\sin(\pi x)/\pi x$. The pulse is clearly NPC; that no narrower pulse has zeros at nT is shown in the text references. The sinc pulse plays a major role in communication theory and will come back in Section 1.3.

Although the zero crossing property was important in Nyquist's time, today it is pulse orthogonality that matters, because such pulses have a simple optimum receiver. But the two concepts are closely related. Equation (1.2) is simply a statement that the autocorrelation of $h(t)$ is itself an NPC pulse. The Fourier transform of this autocorrelation is always $|H(f)|^2$, whether or not h is symmetric. This leads to two properties:

Property 1.3 (**Nyquist Orthogonality Criterion**)   *$h(t)$ is an orthogonal pulse if and only if its autocorrelation function is NPC.*

Property 1.4 *A sufficient condition that $h(t)$ is orthogonal is that $|H(f)|^2$ is antisymmetric about the points $[|H(f)|^2, f] = [|H(0)|^2/2, 1/2T]$ and $[|H(0)|^2/2, -1/2T]$.*

Orthogonal Pulse Examples. An obvious example is the square pulse

$$v(t) = \sqrt{T}, \quad -T/2 < t \le T/2,$$
$$0, \quad \text{otherwise}, \tag{1.4}$$

shown here as a unit energy pulse. Its spectrum is $H(f) = \sqrt{1/T} \operatorname{sinc}(f/T)$. This pulse has very poor spectral properties, since $|H(f)|$ decays only as $\approx 1/fT$, which is far too slowly for applications. The spectrum does not have the spectral antisymmetry (Property 1.4), but $h(t)$ is nonetheless T-orthogonal.

An important practical example is the *root-raised-cosine* (root RC) pulse, so named because its square spectrum $|(H(f)|^2$ obeys Property 1.4 with an antisymmetric piece of a raised-up cosine. The unit-energy time pulse is

$$h(t) = \frac{\sin[\pi(1-\alpha)t/T] + (4\alpha t/T)\cos[\pi(1+\alpha)t/T]}{\sqrt{T}(\pi t/T)[1 - (4\alpha t/T)^2]}, \quad t \ne 0, \ t \ne \pm T/4\alpha;$$

$$(1/\sqrt{T})[1 - \alpha + 4\alpha/\pi], \quad t = 0;$$

$$(\alpha/\sqrt{2T})[(1 + 2/\pi)\sin(\pi/4\alpha) + (1 - 2/\pi)\cos(\pi/4\alpha)], \quad t = \pm T/4\alpha. \tag{1.5}$$

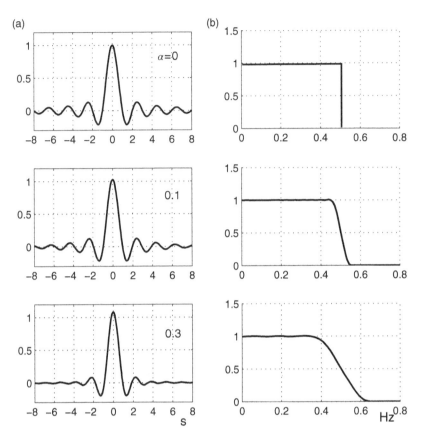

FIGURE 1.1 Root RC orthogonal pulses (a) and their spectra (b) with excess bandwidth $\alpha = 0, 0.1, 0.3$. Unit energy, $T = 1$.

The spectrum is

$$|H(f)|^2 = 1, \qquad 0 \le f \le (1-\alpha)/2T;$$
$$\cos^2[\frac{\pi T}{2\alpha}(f - \frac{1-\alpha}{2T})], \qquad (1-\alpha)/2T < f < (1+\alpha)/2T;$$
$$0, \qquad \text{elsewhere.} \tag{1.6}$$

Here $\alpha \ge 0$, the excess bandwidth factor is the fraction by which the pulse bandwidth exceeds $1/2T$ Hz. Figure 1.1 shows the pulse and spectrum for $\alpha = 0, 0.1, 0.3$, the $\alpha = 0$ case being the sinc pulse. The time main lobe does not differ much but the side lobes rapidly diminish as α grows. The 30% case is arguably the most common pulse in applications, and it will be the standard pulse in most of the book.

The root RC pulse family shows a central fact about pulse design, that bandwidth trades off against time duration. If small interferences with neighboring channels matter, even a small change in bandwidth has a major effect. The uncertainty principle of Fourier analysis states that signal time and bandwidth have a constant product. Beyond this, if ever narrower bandwidth is demanded for a fixed symbol time T, a point must be reached where pulses cannot be orthogonal (it is $\approx 1/2T$ Hz).

The spectra of modulated signals will be taken up in the next section; the error performance and optimum receivers are the subject of Sections 2.1–2.2.

1.3 TIME AND BANDWIDTH

Signal spectra are a crucial issue in this book. Transmission capacity is more sensitive to bandwidth than to energy or signal complexity. Not only width matters but also the *shape* of the spectrum, and particularly, the stop-band side-lobes. The sensitivity heightens as the signal bandwidth efficiency grows. Since bandwidth efficiency is the reason for the book, spectra are a central issue.

The Hertz-Second. Information is of itself timeless. Yet we live in a world where activities are measured by time, and as communication engineers we measure signal bandwidth. In the transmission of information by signals, these resources trade off against each other. For a given parcel of information, if we want to send it faster, we use less time and more bandwidth; if bandwidth is scarce, we take more time. Earlier in the chapter simple modulation signals were defined. By scaling time A-fold faster and power A-fold larger, symbols transmit in the same energy per bit, but are A-fold faster and in A-fold wider bandwidth. If we think in terms of a *joint* time–bandwidth resource, the scaling sets the latency of the transmission but nothing else changes, at least in free space and with sufficient technology. If there are many parcels to send through the same channel, the time–bandwidth available can be divided among the parcels in many ways until the total resource is consumed. The time–frequency consumed by a given transmission will be called its *occupancy* later in the book.

Humans are not timeless, wideband beings with no opinion about latency, but in every application delays up to a limit are acceptable. Thus, some time–bandwidth trade-off is possible and we often accept a time–bandwidth product view. Another compelling reason for such a view is communication theory, which expresses itself most fundamentally and yields its best insight in terms of this product. The view dominates this book. Its unit is the Hertz-second (Hz-s). According to Webster, the Hertz is a " … unit of frequency of a periodic process equal to one cycle per second." A Hz-s is thus dimensionless in the sense that it does not refer to an arbitrarily defined unit like a second. An alien being will see the same quantity that we do.

A fundamental unit of efficiency in communication theory is bits per Hertz-second, abbreviated b/Hz-s. We will call this *bit density* to distinguish it from the more common and more loosely used word *rate*, which can mean input bits per output bit or per second depending on the context. We will avoid common measures of bandwidth efficiency such as bits per Hertz, or its reciprocal Hertz per bit, since these are dimensionally incorrect, and they assume that a second of time has elapsed—they will confuse an alien friend.

Bandwidth Criteria. To form a time–bandwidth product, one must measure bandwidth, but this is not straightforward. To begin with, a finite-time signal has an infinitely wide spectrum. Second, one wishes a single-number measure, whereas the spectrum of a signal is an entire function of frequency. As well, one wants a measure that makes sense with both practical and theoretical signals. Several criteria have evolved and find use in the book.

- *Half-Power Frequency.* Also called 3 dB down frequency; the 3 dB can be another useful value, such as 10 dB. However, because of the spectral anti-symmetry Property 1.1 that applies to orthogonal pulses, the 3 dB version is particularly useful: All reasonable orthogonal pulses with the same symbol time T have the same 3 dB down frequency.

- *Power in Band (PIB).* This refers to the fraction of signal power that lies in the band $[-f, f]$, $f > 0$, expressed usually in percent. The measure applies to baseband signals. For example, the 99% PIB frequency is f inside which lies 99% of the average signal power. The PIB measure is useful because f for 99% or 99.99%, depending on the situation, mark the bandwidths outside of which there will be little interference with neighboring channels. The terms power and energy are used interchangeably, but technically, power applies to an ongoing signal and energy to a single pulse. An alternate measure is the power *out* of band fraction (POB), which refers to the percent outside $[-f, f]$.

- The signal space distance between critical signals can be expressed as a function of frequency and a bandwidth measure derived from that. This method will be taken up in Section 2.5.2.

An analogous problem is measuring the diameter of the sun. The textbook value is 1392680 km, but actually the sun declines in density and has no edge. The size of a kilometer is not in doubt, nor is the Hz, but a density or other feature needs to be specified. In this book, it is most often convenient to use the half-power frequency, because for the important class of orthogonal pulses it has a fixed relation to the symbol time, and the other criteria do not differ much. In some applications and with some pulses, the 99% PIB is more important.

On a mathematical level, all these measures deal in some way with the fundamental problem that a signal or its spectrum or both must lack finite support, while physical signals and spectra clearly have finite support. In an important 1975 paper [11], Slepian discusses this puzzle and asserts that the total bandwidth $\mathcal{W}$ and time $\mathcal{T}$ that one allocates to a signal must always be approximate. Some small fractions ϵ_f and ϵ_t of the energy must lie outside the nominal $\mathcal{W}$ and $\mathcal{T}$. Fixing the fractions defines an occupied bandwidth and time. This is the POB idea.

Figure 1.2 illustrates the concept for a signal that is a sequence of the ten 10% root RC pulses with symbols $[1, -1, -1, -1, 1, -1, 1, -1, -1, 1]$ and $T = 1$. The actual time signal and spectrum are shown. The inner 10 s $\times$ 0.5 Hz box shows the nominal time–bandwidth occupancy of the signal, which is 5 Hz-s. The outer box shows the occupancy that occurs when only fractions $\epsilon_f = \epsilon_t = 0.01$ are allowed outside the

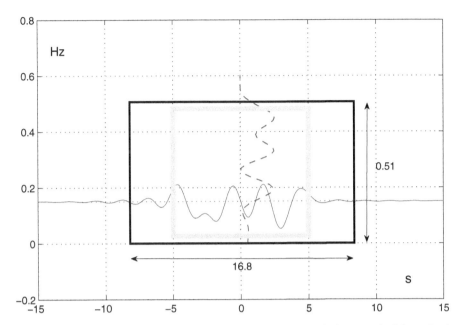

FIGURE 1.2 Time–bandwidth occupancy of a baseband transmission: Nominal (inner box) and actual when only 1% of the pulse energy is allowed out of band (outer box). Ten binary symbols and 10% root RC linear modulation; actual signal (solid line) and spectrum (dashed) are sketched in the boxes.

box.[3] It is clear that the time occupancy now extends far beyond the nominal 10 s, and the occupancy increases to 8.5 Hz-s. For such a short message, a pulse with a better trade-off of frequency and time is needed. There is a time–bandwidth optimization problem here: Given the POB fractions ϵ_t and ϵ_f and a 10-symbol message, what pulse shape minimizes the time–bandwidth occupancy?

We take up occupancy in Chapter 7. The chief conclusions are that the sinc pulse is not optimum and the best pulse is close to a prolate spheroidal wave function.

Signal Spectra. For a linear modulation signal of Eq. (1.1), every set of N bits produces a signal with its own spectrum. Some are wider band than others. They will interfere more with a neighboring channel, or if the channel in use has strict band limitation, they will be damaged more by the limitation. There is a worst case under a given band limitation; this view is taken up in Sections 2.5 and 7.2.2.

The standard view of modulation spectra is to let N grow large and compute the spectrum averaged over the symbol probability distribution. This is called the average power spectral density, denoted as PSD. This need not be the spectrum of

---

[3]The calculation is performed for one 10% pulse, not for the whole waveform; see Chapter 7.

any concrete signal, but it is easily computed and universally used. Much of its convenience stems from the following:

Theorem 1.1 (Linear Modulation Spectrum) *Suppose the same pulse $h(t)$ is used for all N symbols and the symbols u_n are IID with mean zero. Then the PSD is*

$$(1/T)\,\mathcal{E}[|u_1|^2]\,|H(f)|^2. \tag{1.7}$$

That is, the PSD of a linear modulation is the same as the spectrum of its pulse. The proof is simple and available in most digital communication texts ([1], p. 30). Note that the spectrum of the fixed signal in Figure 1.2 is not $|H(f)|^2$ because it is not an average.

The $h(t)$ is primarily a baseband pulse in this book, but the theorem holds if h is a carrier pulse. The IID/mean zero requirement is ordinarily true in practice: If the mean is nonzero, energy will be wasted sending a carrier component; if the symbols are not at least pseudorandom, synchronization will be damaged. When the modulation is coded it is a good assumption, universally applied in this book, that the code symbols are at least uncorrelated, so that the pulse spectrum carries over as well to the coded case.

When the symbols are correlated or the modulation is nonlinear, the PSD is more difficult to compute. Methods exist that are based on cyclostationary random process theory or a random time offset to the signal; the aim is to define a stationary random process with an autocorrelation, which has therefore a spectral density (see [4], p. 61ff). We need the calculation only for CPM coded modulation in Chapter 6.

With the PSD 3 dB bandwidth as a measure, an orthogonal M-ary modulator has time–bandwidth $T(1/2T) = 1/2$ Hz-s, regardless of T and M. The bit density is $2\log_2 M$ b/Hz-s.

The Sinc Pulse. This time pulse has appeared as the Nyquist limit to orthogonal pulses in Section 1.2, and it will appear again in Shannon's capacity calculation in Chapter 3. Controversy surrounds its use, which is worth comment. The sinc and its dual the square pulse are not physically realizable, but to the extent that they can be approximated, they find use. A true time sinc has an attractive spectrum but many disadvantages. Symbol timing cannot be obtained by ordinary means ([1], Section 4.7.2); if symbol timing is not perfect, the sum of the detector ISI is a divergent series. Discrete-time models for faster than Nyquist transmission are difficult to define, and the sinc is far from the solution to Slepian's problem. The heart of the problem is the side lobes in the time domain, which decay only as $1/t$.

The square time pulse has a sinc-shaped spectrum with the dual outcome, that the 1% POB frequency is very large, $\approx 9.5/T$ Hz.

Generally speaking, these outcomes are not acceptable. Sinc pulses are truncated in time, which raises spectral side lobes outside the nominal $[-1/2T, 1/2T]$ Hz bandwidth, and square pulses are truncated in frequency, which creates ISI. The smoother a pulse is in time (the larger root RC α it has) the easier it is to control these effects. Still, even practical smooth pulses can suffer. Truncating a 30% root

RC pulse to time $[-2.5T, 2.5T]$ throws 0.1% of its energy well outside the nominal $[-0.65/T, 0.65/T]$ bandwidth of the pulse ([4], p. 62). The percent may seem small but it is unacceptable interference in many systems.

Wide- and Narrowband Transmission Methods. Whether coded or not, transmission methods are classed as *wideband* if their bit density is less than 2 b/Hz-s. This is because they achieve their rate in b/Hz-s primarily by consuming bandwidth. Everyday examples are space communication and high-quality FM broadcasting. The value 2 b/Hz-s is the density of simple orthogonal pulse binary modulation, as will be shown in Section 2.1. Methods with higher bit density are *high energy*, because they depend chiefly on a high E_b/N_0. These generally need about 3 dB more energy per one-bit increase in density. Short-range wireless links are examples of high energy systems. Note that the ratio E_b/N_0 is what is high, not the bit energy E_b. The choice of wideband or high energy depends on the relative cost of each; there is nothing inherently wrong with either regime.

We will see this distinction in Chapter 2 for modulations, but it is also apparent in Shannon's capacity in Chapter 3.

1.4   CODING VERSUS MODULATION

Not all modulations produce simple pulses and the need to reduce bandwidth can lead to rather complicated signals. It can be a subtle exercise to distinguish coding from modulation, especially when bandwidth plays a role. Some controversy surrounds how to do this. To avoid paradoxes and false hopes, here is a discussion.

Through the development of coding, several concepts have arisen. Coding can be

  (i) the imposition of signal patterns, such as a trellis structure or those imposed by memory;
 (ii) the addition of redundancy, especially through parity check bits;
(iii) the expansion of a signaling alphabet, followed by a selection of words that represent the data, which has a smaller alphabet; and
(iv) selection of a set of some but not all of the possible modulator sequences.

Concept (i) is not suitable for us; we will see that modulators more band limited than orthogonal pulse schemes create trellis-structured signals and require trellis demodulation, even though they are not encoders. Concept (ii) is troublesome because a number of schemes we would like to call coding do not add redundant symbols; one can say that parity check symbols are an artifact of the binary symmetric channel, in which coding can happen no other way. Signal alphabet expansion (iii) is problematical for several reasons: The modulator sets the alphabet, not the coding; the alphabet of some channels, like the AWGN, is the whole real line. Is there a definition of coding wide enough to include all of the schemes we would like to consider?

In his Gaussian channel papers Shannon evolves toward concept (iv). His 1949 paper [7] that introduced Gaussian channel coding concentrates on philosophy, capacity, and his "$2WT$" theorem, which shows that signals over time $\mathcal{T}$ and bandwidth $\mathcal{W}$ span $2\mathcal{W}\mathcal{T}$ orthogonal dimensions. This crucial result is discussed in Section 3.1. It makes possible the capacity theorem and the definition of a code. By 10 years later Shannon would define a code as follows:

> ... a real number may be chosen at the transmitting point. This number is transmitted to the receiving point but is perturbed by an additive Gaussian noise, so that the ith real number, s_i, is received as $s_i + x_i$ ... A *code word* of length n for such a channel is a sequence of numbers $(s_1, s_2, \ldots, s_n)$. This may be thought of geometrically as a point in n-dimensional Euclidean space. ([8], p. 611)

This very nearly captures concept (iv) as it will be implemented in this book. Modulator signals have an expression in terms of orthogonal basis functions, weighted by real numbers. Each number represents a "channel use." The receiver seeks the least-distant whole sequence in Euclidean space. Gallager [9] and Wozencraft and Jacobs [10] in their classic texts essentially concur.

One hesitates to second-guess these authorities, and in any case definition (iv) fits our needs. Earlier, in Section 1.2, the definition of modulator included the idea that *all* M-ary transmission symbols produce outputs. Building on this, we define a channel code as *a set of some but not all of the possible modulator sequences*. If there are N channel uses, the code has rate per use

$$R = (1/N) \log_2(\text{subset cardinality}) \qquad \text{b/channel use}, \tag{1.8}$$

which is R/T in bits/second with a modulator alone. Note that the modulator need not come first in the transmitter, so long as there is some way to connect codewords and modulator outputs.

The rate R must be less than $\log_2 M$. Shannon shows that there is another smaller rate called channel capacity, such that a set and a decoder exist that achieve arbitrarily small error probability as $N \to \infty$.

The codewords in the set can be selected in many ways. Shannon imagined that the letters were chosen at random, and this turned out to be a powerful idea. In Chapter 4, we borrow a convolutional encoder to make the choice; this can be viewed as a pseudorandom selection procedure, and just as Shannon predicted, it works very well. Alternately, words can also be specified on a trellis or graph structure, or as the solutions of equations, as they are in parity-check codes. However the set is selected, some sort of block length N is essential, and it must grow large if R is near capacity.

1.5 A TOUR OF THE BOOK

Chapter 2 introduces the communication theory needed for the book, with emphasis on the issues that play a special role. These include error events, calculation of minimum distance (which predicts the error rate), suitable receiver structures, the

BCJR algorithm (essential in iterative decoding), signal phase (which affects decoding complexity), and the performance of modulators, both simple ones and those producing complicated narrowband signals.

Chapter 3 introduces relevant AWGN-channel Shannon theory. The chapter derives the capacity of this channel and finds from it the Shannon limit to communication as a function of the density (rate) per Hz-s, the signal PSD shape, and the signal energy per data bit. Of these three, density plays the dominant role.

Modulators have associated with them the same three quantities, and an error probability is computed in terms of them in Section 2.1. The simple ones provide a benchmark for low-complexity transmission. A good rule of thumb is that their bit error rates as a function of E_b/N_0 lie about 10 dB from the respective capacity. The performance of coding schemes lies in between these two limits.

Chapter 4 introduces faster than Nyquist signaling (FTN), the most successful method of narrowband coding at present. The term has a 40-year history, and it originally meant an orthogonal pulse modulator with symbol time accelerated; the pulses were no longer orthogonal but the error performance was undiminished. Today FTN means that modulation pulses are nonorthogonal, for whatever reason. FTN methods can be coded, and Chapter 3 shows that they have a better Shannon limit. Chapter 4 explores many aspects that arise in this new technology, including simplified receivers, design of good codes, and error performance analysis.

Classical FTN signals were accelerated in time, but the idea extends to compression of subcarriers in frequency. The outcome occupies less bandwidth and is similar to orthogonal frequency division multiplex (OFDM), but with nonorthogonal subcarriers. Chapter 5 presents this idea and a number of variations. Since OFDM is a favored method in fourth-generation wireless telephony, this "non-O FDM" is attracting interest for fifth-generation systems.

Chapter 6 compares these new methods to older coded modulation methods. These older ideas have bandwidth consumption in between the new ones and binary error-correcting codes. Some FTN implementations in the literature are also reviewed, including chip designs.

Chapter 7 explores alternate ideas about the design of the modulation base pulse itself. One analysis, by Slepian, seeks the pulse with the least time and frequency occupancy. The outcome is related to the IOTA pulse, a popular pulse in OFDM. Another analysis seeks the pulse with the best modulator error performance for a given bandwidth.

1.6 CONCLUSIONS

Where does all this lead? The evidence in this book strongly supports certain conclusions:

Very narrow band energy-efficient transmission cannot occur without *both (i) complicated modulation–demodulation and (ii) significant decoding complexity*. These work together.

Nonorthogonal modulation pulses are necessary. Narrowband transmission is built upon narrowband pulses. Their response is much longer than the data symbol time and it leads to significant ISI.

To perform reasonably near the Shannon limit requires iterative decoding. No other method is available today.

REFERENCES

1. *J.B. Anderson, *Digital Transmission Engineering*, 2nd ed., Wiley–IEEE Press, Piscataway, NJ, 2005.

2. *M. Schwartz, *Information Transmission, Modulation, and Noise*, 4th ed., McGraw-Hill, New York, 1990.

3. *J.G. Proakis, *Digital Communication*, 4th and later eds., McGraw-Hill, New York, 1995.

4. J.B. Anderson and A. Svensson, *Coded Modulation Systems*, Kluwer-Plenum, New York, 2003.

5. H. Nyquist, Certain factors affecting telegraph speed, *Bell Syst. Tech. J.*, pp. 324–346, 1924.

6. H. Nyquist, Certain topics on telegraph transmission theory, *Trans. AIEE*, **47**, pp. 617–644, 1928.

7. C.E. Shannon, Communication in the presence of noise, *Proc. IRE*, **37**, pp. 10–21, 1949; reprinted in *Claude Elwood Shannon: Collected Papers*, Sloane and Wyner, eds, IEEE Press, New York, 1993.

8. Probability of error for optimal codes in a Gaussian channel, *Bell Syst. Tech. J.*, **38**, pp. 611–656, 1959; reprinted in *Claude Elwood Shannon, ibid.*, 1993.

9. R.G. Gallager, *Information Theory and Reliable Communication*, McGraw-Hill, New York, 1968.

10. *J.M. Wozencraft and I.M. Jacobs, *Principles of Communication Engineering*, Wiley, New York, 1965.

11. D. Slepian, On bandwidth, *Proc. IEEE*, **64**, pp. 292–300, 1976.

*References marked with an asterisk are recommended as supplementary reading.

2

COMMUNICATION THEORY FOUNDATION

INTRODUCTION

Signal space theory and the communication theory that stems from it provide the language, methods, and a great many basic results for Gaussian channel transmission. Our purpose in this chapter is to outline the basic ideas, emphasizing those that are important in narrowband transmission. Many good text books are available for the details. Citations will be made to [1], Proakis [2], and the text by Wozencraft and Jacobs [3], which was the first text to present signal space theory and remains the best full-length introduction.

Section 2.1 presents signal space ideas and derives error probabilities of simple modulations. Section 2.2 explores optimal detection under Gaussian noise, beginning with trains of orthogonal pulses and continuing to detection of nonorthogonal pulse modulation. The latter type is the key to narrowband signaling. In the front analog part of the receiver, a simple matched-filter structure turns out to be sufficient for most purposes, and it retains most of the advantages of orthogonal transmission. It forms the input to the trellis detection or BCJR algorithm, and these are discussed next. They depend on discrete-time models, developed in Section 2.4. Faster than Nyquist signaling is introduced at this point as an example of nonorthogonal narrowband modulation. The section concludes with model phase versions, an important concept in theory and implementations. Section 2.5 ends the chapter with an exploration of error events, their signal space distances, and their spectral analysis.

Bandwidth Efficient Coding, First Edition. John B. Anderson.
© 2017 by The Institute of Electrical and Electronics Engineers, Inc. Published 2017 by John Wiley & Sons, Inc.

2.1 SIGNAL SPACE

The fundamental idea of signal space theory is to express a set of continuous-time signals as vectors in a Euclidean space, with a measure of distance between them. Probability of error is a function of this distance, and the components of the vectors are a discrete-time expression of the signals. The first full exposition of the theory appeared in Kotelnikov's 1947 doctoral thesis [5]. Shannon published the theory independently in 1949 [6] as a way to present his AWGN channel results.

The ideas of the theory can be summarized in the points that follow. Transmission occurs via a set of continuous signals $s_1(t), \dots, s_{\mathcal{N}}(t)$ that represent $\mathcal{N}$ messages. Each has an *a priori* probability $P[s_i]$. The Gaussian channel adds a Gaussian random process outcome $\eta(t)$ to produce the output $r(t) = s(t) + \eta(t)$. The process is stationary and white with power spectral density $N_0/2$ W/Hz at all f that are relevant to the discussion.[1] The signals can be the M outcomes of a simple orthogonal modulator pulse, or they may also be a very large, complex set.

- *Maximum Likelihood Receiver.* The optimum receiver seeks the most likely $s_i(t)$ given the known information, that is, it seeks i that maximizes $P[s_i(t)|r(t)]$. With an application of Bayes' rule, $P[s_i(t)|r(t)]$ may be written as

$$P[s_i(t)|r(t)] = \frac{P[r(t) \text{ received } | s_i(t) \text{ sent}]P[s_i(t) \text{ sent}]}{P[r(t) \text{ received}]}.$$

If the *a priori* distribution is known, the receiver performs

$$\text{Find } i \text{ that achieves: } \max_i \ P[s_i(t)|r(t)]$$
$$= \max_i \ P[\eta(t) = r(t) - s_i(t)]P[s_i(t)], \qquad (2.1)$$

since $P[r(t) \text{ received}]$ is a constant during the maximizing. This is called the *maximum aposteriori*, or MAP, receiver. If the *a priori* probabilities are unknown or the uniform distribution, the receiver performs

$$\text{Find } i \text{ that achieves: } \max_i \ P[s_i(t)|r(t)]$$
$$= \max_i \ P[\eta(t) = r(t) - s_i(t)], \qquad (2.2)$$

which is the *maximum likelihood* (ML) receiver.

- *Orthogonal Basis.* Probabilities of the form $P[\eta(t) = r(t) - s_i(t)]$ are not well defined when $\eta(t)$ is a Gaussian random process. The solution to this problem is to construct a set of orthonormal basis functions $\varphi_1(t), \dots, \varphi_J(t)$ for the $\mathcal{N}$ signals. The span of these is the *signal space*. If the signal set is N pulses of an M-ary orthogonal modulation, there are N^M signals in the set and an obvious basis is the set of delayed pulses $\{h(t), h(t - T), \dots, h(t - (N - 1)T)\}$

---

[1]This spectral density is the Fourier transform of the process autocorrelation and is not the PSD of Chapter 1, which is the average of a set of transforms. It is unfortunate that both are called PSDs.

with $J = N$. In other cases, the basis is a subset of the signals which happens to span the entire set. In the general case, the Gram—Schmidt procedure sets up the basis; a proper basis with no more than $\mathcal{N}$ signals is guaranteed to exist ([3], Appendix 7A; [1], Section 2.5). Once the basis is formed, each signal $s_i(t)$ is expressed as the J-component vector $s_i = (s_{i1}, s_{i2}, ..., s_{iJ})$, in which the jth component is the inner product $s_{ij} = \int s_i(t)\varphi_j^*(t)dt$. The time-domain signals can be recovered from

$$s_i(t) = \sum_{j=1}^{J} s_{ij}\varphi_j(t), \qquad \text{all } i. \tag{2.3}$$

The noise waveform $\eta(t)$ and the received $r(t)$ are expressed respectively by $(\eta_1, ..., \eta_J, \eta_{J+1}, ...)$ and $(r_1, ..., r_J, r_{J+1}, ...)$. As with s_i, the components are inner products between the time function and a basis function. Extra dimensions beyond J are shown here because white noise has many more dimensions than the signal set.

- *Theorem of Irrelevance.* Only noise in the J dimensions of the signal set affects the MAP and ML receiver decision. Thus the two standard receivers in vector form become

$$\text{Find } i \text{ that achieves:} \quad \max_i P[\eta = r - s_i]P[s_i] \quad \text{(MAP)} \tag{2.4}$$

$$\text{Find } i \text{ that achieves:} \quad \max_i P[\eta = r - s_i] \quad \text{(ML)} \tag{2.5}$$

in which

$$\eta = (\eta_1, ..., \eta_J)$$
$$r = (r_1, ..., r_J).$$

If the η components are real variables, as they are in the Gaussian case, $P[\ ]$ for η is to be interpreted as a density.

- *Noise Representation.* The components of η and r are ordinary Gaussian random variables. Since it will be needed later, we state this as

Theorem 2.1 *If $\eta(t)$ is a white Gaussian process with spectral density $N_0/2$, its inner products with any set of orthonormal basis functions are IID Gaussian variables $\eta_1, \eta_2, ...$ that satisfy*

(i) $\mathcal{E}\{\eta_j\} = 0$, *all j*

(ii) $\text{cov}(\eta_j, \eta_k) = N_0/2$, $j = k$, *and 0 otherwise.*

This core theorem frees us from the quandaries of stochastic processes and provides the basis for error rate calculations. For a proof see Reference 1, Section 2.5.3; for further discussion of stochastic process issues see a text such as Papoulis [7].

- *Vector Receiver.* Since the η components in Eq. (2.5) are independent Gaussian random variables, the probability density function there is consequently

$$\sqrt{1/(\pi N_0)^J} \ \exp\left[-\sum_{j=1}^{J}(r_j - s_{ij})^2/N_0\right].$$

Taking the log and ignoring constants, we may write the ML receiver as

$$\text{Find } i \text{ that achieves:} \quad \min_i \ \sum_{j=1}^{J}(r_j - s_{ij})^2 = \min_i \ \|r - s_i\| \qquad \text{(ML)} \quad (2.6)$$

and similarly the MAP receiver as

$$\text{Find } i \text{ that achieves:} \quad \min_i \ \left[\|r - s_i\| - N_o \ln P[s_i]\right] \qquad \text{(MAP).} \quad (2.7)$$

Many important facts in receiver design hold true because they trace back to these two lines. Since $\|r - s_i\|$ is the ordinary Euclidean distance between the two vectors r and s_i, the ML receiver works by finding the closest signal to r in Euclidean space. Furthermore, signals can be portrayed as points in the space.

- *Signal Distance.* By Parseval's identity, the form $\|r - s_i\|^2$ has the value $\int [r(t) - s_i(t)]^2 dt$. The square root of either form gives $D(x, y)$, the standard Euclidean measure of distance between x and y. By convention the square distance $D^2(x, y)$ is normalized by $2E_b$, where E_b is the average energy per transmitted symbol, measured bit-wise. The average energy of $\mathcal{N}$ equiprobable signals is $\bar{E} = (1/\mathcal{N}) \sum_i \|s_i\|^2$ or $(1/\mathcal{N}) \sum_i [\int |s_i(t)|^2 dt]$, and they carry $\log_2 \mathcal{N}$ bits, so that the norming factor is $\log_2 \mathcal{N}/2\bar{E}$. This yields the following expression for the *normalized signal space square distance* between two signals:

$$d^2(s_i, s_2) = \frac{\log_2 \mathcal{N}}{(2/\mathcal{N}) \sum_i \|s_i\|} \int [s_i(t) - s_j(t)]^2 \, dt$$

$$= \frac{1}{2E_b} \int [s_i(t) - s_j(t)]^2 \, dt \qquad (2.8)$$

Another convention, shown here, is that lower-case d always denotes normalized distance; as well, calculations are generally with square distance, even though square is often not written.

- *Probability of Error.* The probability $p_2(j|i)$ that $s_i(t)$ was sent, but $s_j(t)$ is closer to $r(t)$, is the same as the probability that one component of η exceeds $D(s_i, s_j)/2$. Because each component is an independent Gaussian, this is

$$p_2(j|i) = 1/\sqrt{\pi N_o} \int_{D/2}^{\infty} \exp(-u^2/N_o) \, du \qquad (2.9)$$

where $D = \|r - s_i\|$. Gaussian integrals in communication are expressed in terms of the tabulated unit-variance zero-mean integral, the Q *function* defined by

$$Q(z) = (1/\sqrt{2\pi}) \int_z^\infty \exp(-u^2/2)\, du. \qquad (2.10)$$

In terms of $Q(\ )$,

$$p_2(j|i) = Q(\sqrt{D^2/2N_0}). \qquad (2.11)$$

or with the normalized d,

$$p_2(j|i) = Q(\sqrt{d^2 E_b/N_0}). \qquad (2.12)$$

- *Minimum Distance.* The two signals that lie closest are the ones most easily confused with one another, and these dominate the transmission probability of error. The quantity

$$d_{\min}^2 = \min_{i,j} (1/2E_b) \int_{-\infty}^\infty |s_i(t) - s_j(t)|^2\, dt, \quad i \neq j \qquad (2.13)$$

is the (square) *minimum distance* of the signal set. An approximation to the overall probability of detection error is $Q(\sqrt{d_{\min}^2 E_b/N_0})$. This is only an estimate but it does have the right exponential behavior at higher E_b/N_0; later we will refine this estimate. Minimum distance is unaffected by translations of the signal set, but the average energy $\bar{E}$ is. Energy is minimized when the centroid $(1/\mathcal{N}) \sum s_i$ is at the origin of the signal space.
- *Receiver Decision Regions.* The error-minimizing receiver is the one that chooses s_i that minimizes $\|r - s_i\|$. This divides signal space into decision regions $S_1, \ldots, S_\mathcal{N}$, one for each s_i. The Euclidean geometry implies many properties; for example, the decision boundary between two adjacent signal points is the perpendicular bisector of the line connecting them.

Simple Modulations. The PAM T-orthogonal modulator family in Section 1.2 is a performance benchmark in this book and provides some initial examples of signal space analysis.

The signal set created by a single-symbol transmission with binary PAM (2PAM) has two signal points $\pm\sqrt{E_s}$, one dimension, and basis function $h(t)$, the same as the unit-energy orthogonal modulation pulse. Since the two signals have difference $2\sqrt{E_s}$ and E_b is E_s, Eqs. (2.8) and (2.12) directly give

$$d_{\min}^2 = (1/2E_b) \int [2\sqrt{E_b}h(t)]^2\, dt = 2, \qquad (2.14)$$

and the true probability of error is $Q(\sqrt{2E_b/N_0})$. The corresponding PSD is $|H(f)|^2$, and measuring bandwidth by the half-power frequency $1/2T$ Hz, we get a bit density of $\log_2 2/[T(1/2T)] = 2$ b/Hz-s, T here being the modulator symbol time. One could instead take an N-pulse set with 2^N signals, but there is no point, because detecting pulse by pulse is an optimal procedure with independent symbols.

Transmission by 4PAM leads to the same scenario except that the space has four points, $\sqrt{E_s/5}\{\pm 3, \pm 1\}$. The four-point signal space is shown in Figure 2.2a, with dashed lines locating the decision boundaries. The minimum distance between any two adjacent points is

$$d_{\min}^2 = (\log_2 4/2E_s) \int [2\sqrt{E_s/5}\, h(t)]^2\, dt = 0.8. \qquad (2.15)$$

This yields the symbol error probability estimate $Q(\sqrt{0.8E_b/N_0})$, which is now only approximate. It will be useful later to have a true computation, and working from first principles, one gets

$$P[\text{Error}|\pm 1 \text{ Sent}] = 2Q(\sqrt{0.8E_b/N_0}),$$
$$P[\text{Error}|\pm 3 \text{ Sent}] = Q(\sqrt{0.8E_b/N_0}).$$

Weighting each case 1/4 for equiprobable symbols gives

$$P[\text{Error}] = (3/2)\, Q(\sqrt{0.8E_b/N_0}) \qquad (\text{4PAM}). \qquad (2.16)$$

The signal PSD is as before, so that the bit density is $\log_2 4/[T(1/2T)] = 4$ b/Hz-s. That is, the bit density is double that of 2PAM, and there is a 2.5-fold energy loss (4.0 dB).

8PAM again leads to the same signal space scenario, but now $d_{\min}^2 = 2/7 \approx 0.286$. The true symbol error rate is

$$P[\text{Error}] = (7/4)\, Q(\sqrt{0.286E_b/N_0}) \qquad (\text{8PAM}) \qquad (2.17)$$

and the bit density is 6 b/Hz-s. This is triple that of 2PAM, with a seven-fold energy loss (8.5 dB).

Multi Dimensional Signal Sets. When there are many signals and more than one dimension, signal space calculations become more complicated. A general form for error probability is

$$p_e = (1/\mathcal{N}) \sum_{i=1}^{\mathcal{N}} P[\mathbf{r} \text{ not in } S_i | \mathbf{s}_i \text{ sent}]. \qquad (2.18)$$

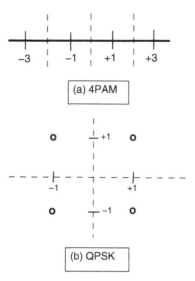

FIGURE 2.1   Signal space constellations for (a) 4PAM and (b) QPSK, showing decision boundaries. For unit symbol energy, divide 4PAM points by $\sqrt{5}$ and QPSK points by $\sqrt{2}$.

An overbound to $P[\mathbf{r} \text{ not in } S_i | \mathbf{s}_i]$ is $\sum_{j \neq i} p_2(j|i)$, with p_2 taken from Eq. (2.11). Placing this in Eq. (2.18) we obtain

$$p_e \leq (1/\mathcal{N}) \sum_{i=1}^{\mathcal{N}} \sum_{j \neq i} Q\left[\frac{\|\mathbf{s}_j - \mathbf{s}_i\|}{\sqrt{2N_o}}\right]. \tag{2.19}$$

This is a sum of $\mathcal{N}(\mathcal{N}-1)$ exponentials that is strongly dominated by those terms that have the minimum distance. Let K be the number of distinct minimum distance terms. The sum counts each term twice so that we can write the estimate

$$p_e \approx (2K/\mathcal{N}) Q(\sqrt{d_{\min}^2 E_b/N_o}). \tag{2.20}$$

This sort of estimate can be quite accurate with narrowband signaling. The E_b/N_0 is higher, which increases the dominance of $d_{\min}$, and the factor $2K/\mathcal{N}$ refines the estimate. In any case, Eq. (2.20) has the correct exponential behavior.

As a second multipoint example, we can estimate p_e for the two-dimensional four-point constellation in Figure 2.1b.[2] There are four distinct point pairs at the square minimum distance; Eq. (2.8) gives the normalized value

$$d_{\min}^2 = \frac{\log_2 4}{(2/4)4E_s}(2\sqrt{E_s}/\sqrt{2})^2 = 2.$$

---

[2] Among other schemes, the constellation describes quaternary phase-shift keying, the most common digital carrier modulation. However, it will not figure explicitly in the book since we employ baseband analysis.

Equation (2.20) thus yields $2Q(\sqrt{2E_b/N_0})$. The true value can be computed by noting that for correct detection of s_3 noise components η_1 and η_2 must both be less than $\sqrt{E_s/2}$. Erroneous detection thus has probability $1 - [1 - Q(\sqrt{2E_b/N_0})]^2 = 2Q(\sqrt{2E_b/N_0}) - Q^2(\sqrt{2E_b/N_0})$. By symmetry, this must be the error for any transmitted signal. Equation (2.20) is thus too high by $Q^2(\sqrt{2E_b/N_0})$, which normally is negligible.

As the book progresses, it will turn out that signal spaces for desirable narrowband schemes are huge, both in dimension and number of points, so that sketches of points and regions are not useful. Signal space insights and particularly minimum distance and related estimates continue to hold true, but different tools are needed. These are based on the trellis structure of bandwidth efficient modulator signals. These tools are taken up in Section 2.5 after models for the modulators are developed.

2.2   OPTIMAL DETECTION

In the practical world, a demodulator has two aims: It converts analog signals to discrete time and it decides M-ary symbols, both, hopefully, in an optimal manner. The structure of this section reflects these aims. Section 2.2.1 applies continuous signals to a matched filter and in so doing converts them to a sequence of discrete-time samples; Section 2.2.2 detects what symbols most likely caused these samples. These jobs need to be done in such a way that the whole procedure is mutually an ML receiver. Section 2.2.1 begins with orthogonal linear modulation, since this is the basis of many coded and uncoded systems, and it provides also the core of a simple approach to nonorthogonal modulation. Nonorthogonal demodulation must deal with intersymbol interference (ISI) as well as additive noise, both of which need to be removed. The signal space approach allows us to convert the essential information to discrete-time values. Section 2.2.2 then decides which symbols were sent. Several methods are developed, including trellis decoding, the BCJR algorithm, and simple equalization.

2.2.1   Orthogonal Modulator Detection

A T-orthogonal pulse satisfies Eq. (1.2). From this alone we can derive a demodulator for the simple linear modulation $s(t) = \sqrt{E_s} \sum u_n h(t - nT)$ and show that it is an ML detector for each symbol u_n. In the end, each u_n will be taken separately, but first consider the entire N-symbol signal. In the absence of noise, Eq. (1.2) implies that $u_n = \int s(\tau)h(\tau - nT)d\tau$. The integral can be written instead as the convolution of $s(\tau)$ and $h(-(\tau))$, evaluated at nT. Adding the white noise process $\eta(t)$ to $s(t)$, we can write the receiver sample as

$$r_n = [s(\tau) + \eta(\tau)] * h(-(\tau))\big|_{nT} = u_n + \eta_n, \tag{2.21}$$

where η_n is a Gaussian noise variate. The successive unit-energy pulses $h(t - nT)$, $n = 0, \ldots, N - 1$ are a valid orthonormal basis for the signal space, and according to Theorem 2.1, $\eta_0, \ldots, \eta_{N-1}$ are IID Gaussians with mean 0 and variance $N_0/2$. The signal space component s_n is directly proportional to $\sqrt{E_s}u_n$ and the ML receiver is given in Eq. (2.6).

Since both the modulator symbols and the noise variates are independent of their neighbors, there is no loss detecting the symbols one at a time. The ML receiver becomes: At time nT find the nearest modulator symbol to r_n. This is implemented by the decision regions in Figure 2.1a. Furthermore, convolution by $h(-(\tau))$ is the same as filtering by transfer function $H^*(f)$. In either case, samples at times $0, T, \ldots, (N - 1)T$ are compared to the noise-free symbol values, and the closest one at nT is the detected u_n. For binary modulation, the values are $\{\pm\sqrt{E_s}\}$; for 4-ary modulation, they are $\{\pm\sqrt{E_s/5}, \pm 3\sqrt{E_s/5}\}$. The modulator/demodulator is equivalent to the diagram in Figure 2.2. This straightforward modulation based on a T-orthogonal $h(t)$ will be called *simple modulation* throughout the book.

Next we investigate the case where $h(t)$ is not T-orthogonal. Nonorthogonal linear modulation can appear in several important ways. In faster than Nyquist signaling (FTN), a T'-orthogonal $h(t)$ is kept fixed while the symbol time T is reduced, so that more pulses are transmitted but $h(t)$ is no longer T-orthogonal; this violates Property 1.2. FTN signaling is the subject of Chapter 4. Nonorthogonality can also come about through filtering. If a filter $G(f)$ is applied to the orthogonal linear modulation $s(t)$—to reduce its bandwidth or because of channel distortions—it is equivalent to a new linear modulation $\sqrt{E_s}\sum u_n w(t - nT)$ with $w = h * g$.

Now time shifts of $h(t)$ cannot be the orthonormal basis functions for signal space. Much of the time the needed basis $\{\varphi_j(t)\}$ can be time shifts of a T-orthogonal function that is no longer $h(t)$ but meets some reasonable conditions. A suitable set can be time shifts of a root RC pulse every nT, like those used in simple modulation; the Nyquist sampling theorem needs to be satisfied so that $s(t)$ can be obtained from its samples. In what follows, we will call such a set a *simple basis*. Signal $s_i(t)$ then has vector representation $(s_{i1}, s_{i2}, \ldots, s_{iJ})$ as usual, with $s_{ij} = \int s_i(t)\varphi_j^*(t)dt$.

But there is a more interesting presentation. The $h(t)$ can be represented as the vector $(\ldots, c_{-1}, c_0, c_1, \ldots)$, where

$$c_n = \int h(t)\varphi_n^* \, dt. \tag{2.22}$$

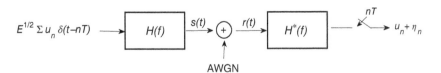

$$E^{1/2} \sum_n u_n \delta(t-nT) \longrightarrow \boxed{H(f)} \xrightarrow{s(t)} \oplus \xrightarrow{r(t)} \boxed{H^*(f)} \longrightarrow \overset{nT}{\diagup} \longrightarrow u_n + \eta_n$$

AWGN

FIGURE 2.2 Linear modulator and matched-filter demodulator.

Let the 0th simple basis function be $v(t)$. Then since $s(t) = \sqrt{E_s} \sum u_n h(t - nT)$ and $h(t) = \sum c_\ell v(t - \ell T)$

$$s(t) = \sqrt{E_s} \sum b_n v(t - nT), \qquad b_n = \sum u_{n-\ell} c_\ell. \tag{2.23}$$

The sequence $\{b_n\}$ is the convolution of the standard M-ary transmission symbols with the sequence $\{c_\ell\}$. In terms of z-transforms, Eq. (2.23) is $S(z) = C(z)U(z).$[3] The new modulation in Eq. (2.23) is the previous orthogonal one, but with a large alphabet of reals replacing the M-ary values u_n and the orthogonal function $v(t)$ replacing $h(t)$. The modulator/demodulator structure continues to be the one in Figure 2.2. It now puts out $b_n + \eta_n$, where η_n as before is an IID Gaussian with mean zero and variance $N_0/2$.

The demodulation is actually incomplete, because it remains to find the symbols u_n that most likely lead to the output, and that is the subject of Section 2.2.2. The setup here works well for most of the schemes in the book, and to distinguish it from others it will be called the *orthonormal simple basis* (OSB) detector.

Given a set of M-ary symbols to be modulated, both Eq. (2.23) and $\sqrt{E_s} \sum u_n h(t - nT)$ produce the same $s(t)$. Most often, the nonorthogonal $h(t)$ is designed to reduce bandwidth, the sampling theorem is satisfied, and Eq. (2.23) can be viewed as accomplishing this in discrete time.

With most basis functions that one might choose, the sequence $\{c_n\}$ is directly the samples of $h(t)$ at nT. When this is actually the case is given by

Property 2.1 (**OSB Condition**) *Let $H(f) = 0$, $f > W$, with $W < 1/2T$ Hz, and let the 0th pulse $v(t)$ of a simple basis set satisfy $V(f) = C_0$ a constant for $|f| < W$. Then $\{c_n\}$ are the samples $h(nT)$ and Eq. (2.23) reproduces all modulated signals.*

The proof uses Fourier transform properties (see Reference 8, Section 6.2.1). For analysis purposes, the identity of $v(t)$ is of no consequence so long as $v(t)$ satisfies the property; in an implementation along the lines of Figure 2.2, the choice sets the analog-domain filter $V^*(f)$, the front-end matched filter in the receiver. When the condition in the Property fails, signal $s(t)$ cannot be reproduced from its samples each nT. There exist examples where the detection is suboptimum. However, the motivation for $h(t)$ is usually its narrow bandwidth, so that the condition holds.

Example 2.1 *(Sampling a Non-orthogonal $h(t)$)*
Figure 2.3 shows an example where $h(t)$ is the unit-energy 30% root RC pulse that is orthogonal with reference to $1.5T$, $T = 1$, and is constructed from nine basis functions, where $v(t)$ is a 12% T-orthonormal root RC pulse. The inner products $\{c_\ell\}$ work out to be the samples $h(nT)$. The error in the approximation (shown

---

[3]The z-transform of a time-discrete sequence $y_0, y_1, y_2, \ldots, y_K$ is $Y(z) = y_0 + y_1 z^{-1} + y_2 z^{-2} + \cdots + y_K z^{-K}$. Convolution in time is multiplication in the z-domain.

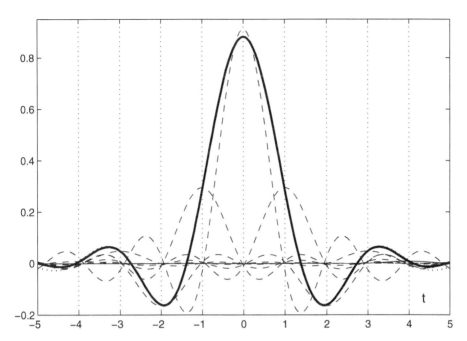

FIGURE 2.3 Construction of $h(t)$ from a simple orthogonal basis. Nine 12% root RC 1-orthogonal pulses centered on $-4, -3, \ldots, +4$ approximate a 30% 1.5-orthogonal root RC pulse. Heavy curve is the approximation; dots show true $h(t)$.

as dots) is only visible at $|t| > 4$. When $h(t)$ is a 60% root RC pulse, $h(t)$ fails Property 2.1; $\{c_1, c_2, c_3\}$ are $\{-0.869, -1.318, -0.101\}$, compared to the $h(t)$ samples $\{-0.884, -1.329, -0.080\}$; a slight approximation error appears as a ripple, caused by aliasing (see Section 2.3). This error may cause no measurable difference in detection.

The idea of convolving the symbols $\{u_n\}$ with a fixed sequence, the "generator," has been with us since the 1960s, mostly under the name partial response signaling (PRS). Much can be done with this technique, and for a general view see Reference 8, Chapter 6. Some history of PRS is reviewed in Chapter 4.

Some Other Receiver Structures.

Correlator Receiver. Starting from the definition of the ML receiver Eq. (2.6), and applying Parseval, we have

$$
\begin{aligned}
||r - s_i||^2 &= \int |r(t) - s_i(t)|^2 \mathrm{d}t \\
&= \int |r(t)|^2 \mathrm{d}t + \int |s_i(t)|^2 \mathrm{d}t - 2 \int r(t) s_i(t)\, \mathrm{d}t.
\end{aligned}
$$

Minimizing the left-hand side here over i is the same as finding

$$\max_i \int r(t)s_i(t)\,dt - \varepsilon_i/2, \tag{2.24}$$

in which ε_i is the energy of the ith signal. If all signals have the same energy, ε_i can be ignored. The integral is just the correlation of $r(t)$ with signal i, and Eq. (2.24) is the correlation receiver.

Matched Filter Receivers. Equation (2.24) can be viewed as linear filtering. That is, if $g(\tau) = s_i(-\tau)$ is the filter impulse response, the correlation is

$$\int r(\tau)s_i(\tau)d\tau = \int r(\tau)g(t-\tau)d\tau\Big|_{t=0} = r * g\Big|_{t=0}, \tag{2.25}$$

the filter output at time 0. Note that $G(f) = S_i^*(f)$, and the receiver structure is again the receiver side of Figure 2.2, with $S_i^*(f)$ replacing $H^*(f)$. Taking the max over i in Eq. (2.25) is the matched filter receiver in its classical version, acting on entire signals $s_i(t)$.[4] Once again, energies need to be subtracted if they differ.

In another manifestation of the matched filter receiver, $r(t)$ is matched-filtered by the full nonorthogonal $h(t)$; that is, not by a signal $s_i(t)$ and not by an orthogonal pulse as in Eq. (2.21). The entire modulator/demodulator model is precisely Figure 2.2, with the new $h(t)$. The remainder of the receiver works with the samples of the filtering at kT, $k = 0, 1, \ldots$, which are given by

$$w_k = r(\tau) * h(-\tau)\Big|_{t=kT} = \left[\eta(\tau) + \sqrt{E_s}\sum_{n=0}^{N-1} u_n h(\tau - nT)\right] * h(-\tau)\Big|_{t=kT}$$

$$= \xi_k + \sqrt{E_s}\sum u_k \rho_h[k - n]. \tag{2.26}$$

Here, $\rho_h[k - n]$ is the autocorrelation of $h(t)$ at $(k - n)T$ and $\xi_k = \eta(\tau) * h(-\tau)|_{t=kT}$ is a sample of filtered white noise.[5] As z-transforms this is

$$W(z) = U(z)R_h(z) + \Xi(z), \tag{2.27}$$

where $R_h(z)$ is the z-transform of $\rho_h[k]$. Since $h(t)$ is not T-orthogonal, $\{\xi_k\}$ is a *colored* noise sequence, which means that distance in the subsequent processing is not Euclidean. This receiver was developed by a number of researchers in the 1970s.

Whitened Matched-Filter (WMF) Receiver. A so-called spectral factorization of $R_h(z)$ resolves the colored noise problem at least some of the time. The procedure will be taken up in Section 2.4, along with a proof that the WMF method has optimal

---

[4]The filter as stated is noncausal. In a practical receiver, $g(\tau)$ is delayed a suitable KT so that $g(\tau)$ is small, $\tau < 0$, and the convolution is sampled at time KT. The same procedure needs to be followed with Eq. (2.21) and with the WMF receiver to follow.

[5]Here and throughout, the bracket form $[k]$ denotes the kth value of a sequence.

error performance, but at this point we can quote two of the outcomes. The transform $R_h(z)$ can be factored as $R_h(z) = G(z)G^*(z^{-1})$ such that $1/G^*(z^{-1})$ is a stable filter and $N(z) = \Xi(z)/G^*(z^{-1})$ represents zero-mean white noise with variance $N_0/2$; second, the distance structure of the signal set represented by $U(z)R_h(z)/G^*(z^{-1})$ is the same as that of the original signal set $\{\sum u_n h(t - nT)\}$. This means that the outputs with transform

$$U(z)R_h(z)/G^*(z^{-1}) + N(z) \qquad (2.28)$$

represents the same type of detection problem as Eq. (2.23). The filter $1/G^*(z^{-1})$ is called a whitening filter. The WMF receiver originated with Forney[9], who applied the earlier technique of whitening to symbol detection in ISI.

The WMF is an attractive alternative to the potentially complicated max operations in the correlation and matched filter receivers, but it has fallen somewhat out of favor. The main reason is that the whitening filter, $1/G^*(z^{-1})$, is not stable when $h(t)$ has spectral zeros, which in recent years is an important case; the process of approximating $h(t)$ so that the outcome has a stable whitener is not straightforward. A second reason is that the physical analog matched filter is not root RC or another standard model, but is a more complicated response $h(t)$ that may change. The OSB design, on the other hand, makes it clear how to approximate a troublesome $h(t)$, has no stability problems, and relates to a simpler matched filter.

Gram-Schmidt Receiver. When the signal set is well defined, the set of basis functions $\{\varphi_j(t)\}$ in Eq. (2.3) can be computed and a set of filters constructed that are matched to them. Since the filter responses are orthogonal, the noise variates in their samples will be IID Gaussians. This kind of receiver has been suggested by Kanaras, Darwazeh *et al.* for use with FFT-based frequency FTN; see Section 5.3. Problems have been observed with the stability of matrices used in this method.

Fractional Sampling Receivers. When bandwidths are such that symbol-time sampling errors occur, an alternative is to sample twice or more per symbol time, a method called fractional sampling. With many samples per symbol—3–5 are probably enough—the receiver essentially works with continuous signals.

2.2.2 Trellis Detection

When the output of the front of a demodulator has the Eq. (2.23), a more advanced detection is required for ML detection. Equation (2.23) means that the presented discrete-time sequence is a convolution of the desired symbols $\{u_n\}$ with a generator $\{c_\ell\}$. This convolution form applies with ISI, with channel filtering, and for a different number system, with convolutional coding. The optimal detection structure in all these cases is the *trellis detector*: The trellis detector finds the nearest sequence $\{b_n\}$, with $b_n = \sum u_{n-\ell}c_\ell$, to the input sequence $\{r_n\}$. In more fundamental terms, it finds the nearest outcome of a Markov process.

The detector searches systematically through a plot of the underlying Markov states versus time. The states in our demodulator application are the last μ values of $\{u_n\}$ up to and including the present time $n_0 T$. The plot is called a *trellis*. If u_n is M-ary, there are M state "paths" out of each stage-n_0 state; these run through the trellis and terminate in a state at a later stage; M paths terminate in each state at each stage. A link from one state to the next in the path is a "branch," and associated with each branch is a value, or "label." With an OSB receiver the label is a b_n. Specifically, the label at stage $n_0 + 1$, given the state $u_{n_0 - \mu + 1}, \dots, u_{n_0}$, is

$$b_{n_0 + 1} = \sum_{\ell=1}^{\mu} u_{n_0 - \ell + 1} c_\ell + u_{n_0 + 1} c_0.$$

During detector operation the incoming value $r_{n_0 + 1}$ is compared to this, and the increment $(r_{n_0 + 1} - b_{n_0 + 1})^2$ is added to the Euclidean distance of the path accumulated so far up to time n_0. The key to trellis detection is the trellis Principle of Optimality: When two or more paths reach the same state at time n_0 (called a "merge"), all but one that is closest to the incoming sequence $\dots, r_{n_0 - 1}, r_{n_0}$ can be dropped from further consideration. The step of adding the increment, comparing paths, and dropping all but one is called an add–compare–select. When the algorithm reaches the end of $\{r_n\}$, it will have identified at least one most likely whole sequence $\{u_n\}$ out of the set of possible sequences.

The common name for this algorithm type is Viterbi algorithm (VA), after Andrew Viterbi, who first cited the Principle of Optimality in a 1967 paper on convolutional codes. We call it trellis detection only to distinguish the usual application to convolutional codes from the application here to demodulation. We assume the reader has some familiarity with this well-known scheme.

The VA need not retain path memories indefinitely, but can instead release their symbols as output after an observation window of width $L_D > \mu$, the *decision depth*; L_D is a function of the generator $\{c_\ell\}$ and can be computed by a suitable algorithm based on a trellis search [8]. There are M^μ states, potentially a large number. This motivates schemes that reduce the states, hopefully with only slight loss of ML detection performance. Once c_ℓ falls to a small value, the remaining values $c_{\mu+1}, c_{\mu+2}, \dots$ can be ignored since they make insignificant contribution to the distance increment. In bandwidth- or energy-efficient transmission methods, however, this is often not enough, and M^μ is still uncomfortably large.

If more reduction is needed, there are several strategies to follow. In *reduced search* detection, the VA exhaustive search of the trellis is limited to those areas where $\{b_n\}$ is observed to be reasonably close to $\{r_n\}$. Sequential decoding and the M-algorithm are examples (see, e.g., Reference 8, Chapters 5 and 6). It is known that large reductions in computation are possible by these methods, and they will be illustrated later in the book. In *reduced trellis* detection, also called channel shortening, the generator is shortened but the full trellis is searched; the idea is to give up some error performance in return for a much smaller trellis. Yet another strategy is to split the decision depth into a more recent section where the full identity of all paths is kept and an earlier section where only a single identity is kept. This works because the contribution of earlier symbols to the branch labels is small. Both computation and storage are reduced.

We have only discussed trellis detection with reference to a discrete-time incoming sequence, and virtually all applications are of this sort. But the method can actually be applied to the original set of continuous functions, which contain the same Markov property as do the time-discrete $\{r_n\}$. When it is unclear how to design a matched-filter/sampler receiver, this can be the safest procedure. Indeed, trellis detection adapted to continuous signals is the ultimate receiver, and its minimum distance and error performance are the benchmark for all others.

Simple Equalizers. Equalizer is an older term for a simple method, usually a filter, designed to remove ISI. If the generator $\{c_\ell\}$ is simple and short enough, or if trellis decoding is too complex, an equalizer can be an attractive replacement. The simplest equalizers are linear filters designed to cancel ISI (zero forcing) or minimize noise and/or ISI (minimum square error equalizers). Some include a feedback of tentative symbol decisions (feedback equalizers). In general, equalizers become ineffective as the zeros of $C(z)$ near the z-plane unit circle. Trellis detection is then necessary.

Successive Interference Cancellation (SIC). Iterative receivers will play an important role in Chapters 4 and 5. In these, two processors work on the decoding and feed information to each other; for example, one can be a trellis detector or equalizer and the other a convolutional decoder. In SIC, the first processor removes as much intersymbol or other interference as it can by relatively simple methods, and passes the signal to the second processor. The second may be an error-correcting decoder or another equalizer, for example, one that works in frequency instead of time. In the decoder case, it decodes as well as it can, and passes an improved estimate of the interference to the first processor, which *subtracts* the estimate. It then tries to estimate the remaining interference, and the cycle repeats; the iterations continue until some sort of convergence. The processors may pass soft information, the SIC may subtract soft estimates, and it may be quite suboptimal (to reduce complexity). SIC receivers play an important role in Chapter 5, where equalization of interference is required in two dimensions: frequency and time.

Sphere Detectors. This receiver employs a search over a nearby sphere of possibilities. Representative papers on the general method are References 11 and 12. An application of the Gram–Schmidt and sphere techniques to frequency FTN (Chapter 5) is Reference 14.

The BCJR Algorithm. This scheme is a "soft" trellis decoder that computes symbol or trellis branch *probabilities*. It can also accept *a priori* symbol probabilities if they are known, and in this case it is an MAP detector like Eq. (2.7), but for individual symbols rather than entire signals. It resembles the trellis detector in that it assumes a Markov input and is based on a trellis structure, but here the resemblance ends. Instead of an add–compare–select, the algorithm is based on two linear recursions. By observing its probability outputs, one can make symbol decisions, but the importance of BCJR is chiefly its use in iterative decoding, where the certainty of a decision builds up over many iterations.

The algorithm was devised in the early 1970s by researchers in iterative decoding and first appeared in a 1974 paper by Bahl, Cocke, Jelinek, and Raviv [15], after whom the algorithm is named.[6] It builds upon an earlier algorithm by Baum and Petrie for identifying Markov models; its notation comes from the earlier work. Because the literature contains few explanations of the algorithm, it is described in some detail here. A vector approach is taken, first given in Reference 17. Primarily, the vectors yield a much more streamlined presentation, but concepts such as eigenvectors have physical meaning. More BCJR details appear in References 8 and 16.

The heart of the BCJR algorithm is the calculation of two working vectors called α_k and β_k at each trellis stage k. At each k, the components $\alpha[0], \alpha[1], \ldots, \alpha[S-1]$ and $\beta[0], \beta[1], \ldots, \beta[S-1]$ are special probabilities for each state in an S-state trellis. The kth row vector α_k is defined by

$$\alpha_k[j] \triangleq P\big[\text{Observe } r(1 : k), \text{ Generator in state } j \text{ at time } k\big]. \tag{2.29}$$

The notation $r(1 : k)$ means the incoming values $r_1, \ldots, r_k$. Each r_k is the observation at one stage; if a convolutional code places q bits on each branch, r_k comprises q values. "Generator" refers to the Markov process (convolution, usually) that generates incoming values before noise, in this book either the $\{b_n\}$ or the branch symbols of a convolutional codeword.

The column vector β_k at stage k is defined by

$$\beta_k[i] \triangleq P[\text{Observe } r(k+1 : K) \mid \text{Generator in state } i \text{ at time } k]. \tag{2.30}$$

We also need a matrix Γ_k at each stage k, whose i, j element is

$$\Gamma_k[i, j] \triangleq P[\text{Observe } r(k), \text{ Generator in } j \text{ at time } k \mid \text{Generator in } i \text{ at } k-1]. \tag{2.31}$$

When AWGN adds to a single-branch label b_k, as it would with Eq. (2.23), the elements of Γ_k are simply

$$\Gamma_k[i, j] = \begin{cases} P[u'](1/\sqrt{\pi N_0})\exp(-(r_k - b'_k)^2), & \text{if transition } i \to j \text{ exists} \\ 0, & \text{otherwise} \end{cases} \tag{2.32}$$

where b'_k is a branch label, u' is the modulation symbol corresponding to $i \to j$, and $P[u']$ is its *a priori* probability. If none is known, $P[u']$ is set to 1. In the case of convolutional decoding with several channel symbols on a trellis branch, the Gaussian exponential is replaced with the product of several.

---

[6]In part because the algorithm was too complex for 1970s computing, this well-written paper was mostly ignored until the 1990s, when the BCJR became a critical element in turbo coding. Only then did it win a paper prize in 1998.

Note that $\alpha_k[j]$ is the probability of the incoming channel observations at times $1, \ldots, k$ and at the same time state j at time k; $\beta_k[j]$ is the probability of the observations at $k+1, \ldots, K$ given state j at k; Γ_k is a modification of the usual state transition matrix to include the outcome r_k. The algorithm seeks the probabilities of states *given* the entire incoming $\{r_n\}$, but since $\{r_n\}$ is fixed throughout, the algorithm need only find the probabilities of the states *and* $\{r_n\}$.

Next we define the two linear recursions. We assume that the generation starts and ends at a known state and there are K observations. The *forward recursion* of the BCJR algorithm is

$$\alpha_k = \alpha_{k-1}\Gamma_k, \qquad k = 1, \ldots, K \qquad (2.33)$$

with $\alpha_0[i]$ set to 1 for the known start state i and 0 otherwise. The *backward recursion* is

$$\beta_k = \Gamma_{k+1}\beta_{k+1}, \qquad k = K - 1, \ldots, 1 \qquad (2.34)$$

with $\beta_K[j]$ set similarly to 1 for the known end state. The object now is to find the probability of the transition from state i to state j during stage k, having observed the entire incoming $\{r_n\}$. This is given the notation

$$\sigma_k[i \to j] = P\big[\,\{r_n\}, \text{ Generator in state } j \text{ at } k, \text{ and state } i \text{ at } k - 1\big]. \qquad (2.35)$$

Then it can be shown [8,15] that

$$\sigma_k[i \to j] = \alpha_{k-1}[i]\,\Gamma_k[i,j]\,\beta_k[j]. \qquad (2.36)$$

This is the core calculation of the BCJR algorithm. The proof requires some careful Markov probability calculations; indeed, it can be said that the BCJR consists of simple recursions that are not at all simple to explain.

The basic BCJR thus finds the trellis transition, or branch, probabilities, not the probabilities of the driving symbols. The probability that $\{r_n\}$ occurs and the encoder reaches state j at time k is, using Eq. (2.36),

$$\sum_i \sigma_k[i \to j] = \sum_i \alpha_{k-1}[i]\,\Gamma_k[i,j]\,\beta_k[j] = \alpha_k[j]\beta_k[j], \quad \text{all } j. \qquad (2.37)$$

This is the component-wise product of vectors α_k and β_k and in BCJR analysis it is given the name λ_k. The vector λ_k lists all the state node probabilities at stage k; it is usually normalized to unit sum at each k, which removes the dependence on $\{r_n\}$.

It remains to find the probability of a given symbol at stage k. This is the sum of the probabilities of all transitions that imply this symbol,

$$P[u_n = v] = \sum_{i,j \text{ in } \mathcal{L}} \sum \sigma_k[i \to j] \bigg/ \sum_{\text{all } i,j} \sum \sigma_k[i \to j] \qquad (2.38)$$

Here, $\mathcal{L}$ is the set of trellis transitions that correspond to value v. With demodulation under ISI v is one of the M-ary modulation symbols; in convolutional decoding it is a data symbol, which we assume to be 1 or 0. The denominator term normalizes the outcome to unit sum.

This then is a brief derivation of the BCJR algorithm. What follows are some important comments and extensions that have implications in the later chapters.

- *Storage Demands.* The algorithm needs to retain all α but not all β. Having found all the α_k, it can execute the β recursion one stage at a time, find β_k, σ_k, and λ_k and the desired probabilities at stage k, and then drop the α_{k+1} and β_{k+1} that it just used.

- *Short α-Block BCJR.* The α can be found in short blocks as needed, proceeding backward through the trellis, with each short α-block discarded after use. This avoids storage of the entire α set. Calculation of a short block must start from a guessed initial α_k, but this works well because the recursion settling time to an accurate α is only $\approx L_D$, the decision depth of the generator [18].

- *Use of A Priori Probabilities.* If available, these are used to compute all the α and β. Equation (2.36) then performs the core calculation that finds $\sigma_k[i \rightarrow j]$. It employs the α before k and the β after k, but the *a priori* probabilities *at* stage k only in the factor $\Gamma_k[i, j]$. When there are *a priori* probabilities, $\sigma_k[i \rightarrow j]$ leads to a MAP computational outcome.

- *One-Shot MAP Versus Iterative BCJR Application.* In iterative detection, two BCJRs supply each other's apriori information in a loop, eventually converging to a joint solution. For example, one BCJR may treat ISI, while the other decodes a convolutional code. In such an application, the incoming *a priori* probabilities for stage k cannot be used to compute the output for k, because the stronger of the two BCJRs will then dominate the loop and force a suboptimal convergence. Accordingly, the *a priori* contribution $P[u']$ at k to $\Gamma_k[i, j]$ is removed, by setting it to a constant (normalizations applied in Eqs. (2.37) and (2.38) mean the constant is arbitrary). *The iterating BCJR outcome is no longer MAP.*

- *Log-Likelihoods.* In implementations, probabilities are most often exchanged between BCJRs as log-likehood ratios (LLRs). For an M-ary symbol u, taking values $a_1, \ldots, a_M$, the LLR(u) is an M-tuple whose m-th value is

$$\ln \left(\frac{P[u = a_m]}{P[u \neq a_m]} \right). \tag{2.39}$$

With binary symbols, this takes the simple form

$$\text{LLR}(u) = \ln \left(\frac{P[u = +1]}{P[u = -1]} \right). \tag{2.40}$$

Simply checking the sign of Eq. (2.40) decides the bit u. These forms are a compact way to exchange information, but conversions back and forth are required since the BCJR uses probabilities in its calculations, not LLRs.

- *Max-Log-MAP Algorithms.* Some of the work of passing in and out of logarithms can be relieved by approximating logs by a piecewise linear function. The best known of these methods is the Max-Log-MAP algorithm proposed by Robertson et al. [13]. All such algorithms exact a performance penalty.

- *Precision Problems.* During recursions, α_k and β_k continually decline in value and need to be normalized periodically to unit sum at a stage. Because of the later norming of Eqs. (2.37) and (2.38), this does not affect the outcome. As well, probabilities very near 1 cannot be adequately represented by any reasonable number system (since $1 + \epsilon$ becomes 1 for small ϵ). One solution to this is to retain all M LLR components, even in the binary case. Alternately, approximate BCJRs exist that handle the problem.

- *Colored Noise.* The BCJR algorithm has been adapted to the case of colored noise, such as occurs with the matched filter receiver Eq. (2.27). A recent paper is Reference 19.

2.3   PULSE ALIASING

When the bandwidth of a pulse exceeds $1/2T$ Hz, and the signal waveform is sampled each T, the pulse is said to suffer aliasing. The portion of the pulse spectrum above $1/2T$ folds back about $1/2T$ and adds to the portion below. Since sampling is a fundamental part of receivers that work in discrete time, aliasing is an important issue. Aliasing is especially important in the WMF receiver and the capacity calculation for linear modulation in Section 3.2.

For our purposes, pulse aliasing occurs if and only if the pulse spectrum $|H(f)|^2$ fails to equal the *folded spectrum* $|H_{\text{fold}}(f)|^2$ in the range $[-1/2T, 1/2T]$ Hz. The folded spectrum appears often in communication theory and is defined by

$$|H_{\text{fold}}(f)|^2 = \sum_{n=-\infty}^{\infty} |H(f + n/T)|^2, \quad \text{all } f. \tag{2.41}$$

It is a superposition of replicas of $|H(f)|^2$ centered on all multiples of $1/T$ Hz. Clearly, aliasing cannot occur if $|H(f)| = 0, f \geq 1/2T$, because replicas cannot interfere with each other. The Eq. (2.41) holds for all f, although the structure of H_{fold} is such that the function is defined by its behavior on $[0, 1/2T]$ and we will often refer only to that.

The various degrees of pulse aliasing and how they relate to orthogonality are shown in the folded spectra of Figure 2.4. In each case several spectral replicas are shown and their sum, the folded spectrum, is a heavy line. The axes are multiples of T versus multiples of $1/T$ Hz (alternately, take $T = 1$). Figure 2.4a illustrates the no-aliasing case: The pulse spectrum does not exceed $1/2T$ and no replicas overlap. But the pulse fails Nyquist's spectral antisymmetry condition (Property 1.1) and it is not T-orthogonal. The pulse spectrum in Figure 2.4b does satisfy the antisymmetry condition, and because of the condition, it and its replicas add to the constant value T. In fact, the condition $|H_{\text{fold}}(f)|^2 = T$, a constant, is necessary and sufficient for pulse T-orthogonality; it was published long after Nyquist in 1965 [20] and is called

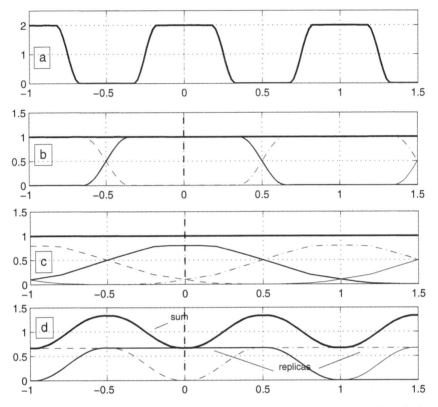

FIGURE 2.4 Four folded spectrum cases. (a) No aliasing, not orthogonal; (b) Aliasing, orthogonal with antisymmetry; (c) Aliasing, orthogonal without antisymmetry; (d) Aliasing, not orthogonal. The heavy line in (a)–(d) is the sum of all replicas. Vertical axis in multiples of T seconds; horizontal axis in multiples of $1/T$ Hz.

the Gibby–Smith (GS) condition. Figure 2.4c shows an example of an aliasing pulse of width 2×1.3 Hz that satisfies GS but not the antisymmetry condition. Figure 2.4d shows an aliasing pulse that is not GS and therefore not T-orthogonal. Spectra a, b, and d are frequency scalings of a standard 30% RC spectrum by a factor 0.5, 1, and 1.5, but only b is orthogonal.

The WMF receiver in Section 2.2.1 samples its matched filter output each T and is well-defined only when the pulse bandwidth exceeds $1/2T$ Hz. Thus, $H_{\text{fold}}(f)$ plays a major role in its operation. We saw that the z-transform of the samples is Eq. (2.27), $U(z)R_h(z)$ plus noise, where $R_h(z)$ is the z-transform of the discrete-time pulse autocorrelation, $\rho_h[k]$. We can define the Fourier transform of $\rho_h[k]$ to be

$$R_{\text{dt,h}}(f) = R_h(e^{j2\pi fTk}) = \sum_k \rho_h[k]e^{-j2\pi fTk}, \qquad (2.42)$$

where the subscript dt signifies discrete time. By the Poisson sum formula of Fourier analysis, the right side is $(1/T) \sum_n R_h(f + n/T)$, in which

$$R_h(f) = \int_{-\infty}^{\infty} \rho(t) e^{-j2\pi f t} \, dt, \qquad (2.43)$$

which is the continuous transform of the full $\rho_h(t)$. From autocorrelation properties this must be $|H(f)|^2$. What we have shown, therefore, is that

$$R_{dt,h}(f) = (1/T)|H_{fold}(f)|^2, \quad \text{all } f. \qquad (2.44)$$

The WMF receiver acts in a sense as if the modulation pulse is aliased, and combats the aliasing. The OSB receiver, on the other hand, works with nonaliased signals, for which $R_{dt,n}(f) = (1/T)R_h(f) = (1/T)|H(f)|^2$, $|f| < 1/2T$. There is no effective difference between continuous and discrete time.

2.4   SIGNAL PHASES AND CHANNEL MODELS

The sampled matched filters in Section 2.2 convert continuous signals to discrete time, and in so doing they also create a Markov model of the signal. In the OSB case, the model is the convolution $b_n = \sum u_{n-\ell} c_\ell$ with the memory-μ generator, or "taps", $c_0, \dots, c_\mu$ that are samples of the base modulator pulse. In the WMF case, a different generator will arise, providing the sample autocorrelation z-transform has no unit-circle zeros. In what follows, we will see that this second model is still a convolution and has the correct distance structure. In fact, models related to a given signal set form groups whose members are phase versions of each other. All these versions are coupled to AWGN noise variates $\eta_1, \eta_2, \dots$. Yet more versions can be produced that are coupled to colored noise, but we will not pursue them.

We first calculate the minimum distance of a channel model. Once obtained, a model has a minimum distance d_{min}^2 and a symbol error probability whose exponential behavior is the same as $Q(\sqrt{d_{min}^2 E_b/N_0})$. With a proper model, the integral form in Eq. (2.13) converts by Parseval's theorem to

$$d_{min}^2 = \min_{i,j} (1/2E_b)\|s_i - s_j\|^2, \quad i \neq j$$

$$= \min_{i,j} (1/2E_b) \sum_n \left[\sum_{\ell=0}^{\mu} u_{n-\ell}^{(i)} c_\ell - u_{n-\ell}^{(j)} c_\ell \right]^2$$

$$= \min_{i,j} (1/2E_b) \sum_n \left[\sum_{\ell=0}^{\mu} \Delta u_{n-\ell} c_\ell \right]^2. \qquad (2.45)$$

Here, Δu denotes the difference $u^{(i)} - u^{(j)}$ and $\sum_n$ runs over all n for which a $\Delta u_{n-\ell}$ is nonzero. The point here is that because convolution is linear only symbol differences matter in a distance calculation.

Calculation of minimum distance is quickly done with modern computers, and is an important tool in signal analysis. Since the algorithms are rather technical, they are relegated to Appendix 2A. Routines are given there for the important cases in this book, which are binary and 4-ary trellis demodulation. In the binary case, the symbols are $\pm\sqrt{E_s}$, $E_b = E_s$, and Eq. (2.45) reduces to minimizing over a set of Δu sequences the expression

$$(1/2) \sum_n \left[\sum \Delta u_{n-\ell} c_\ell \right]^2, \qquad (2.46)$$

where the Δu take any of the values $\{0, \pm 2\}$. The sequence Δu is called an *error difference sequence*. In the 4-ary case, the symbols are $\{\pm\sqrt{E_s/5}, \pm 3\sqrt{E_s/5}\}$, $E_b = E_s/2$, and Eq. (2.45) reduces to minimizing

$$(1/5) \sum_n \left[\sum \Delta u_{n-\ell} c_\ell \right]^2 \qquad (2.47)$$

where the Δu take the values $\{0, \pm 2, \pm 4, \pm 6\}$.

As an illustration of distance calculation, take the simple generator 6, 3, 3, 2, which after normalizing is $c = [0.7878, 0.3939, 0.3939, 0.2629]$ (henceforth generators are given in vector notation). This could represent intentional filtering or ISI, among other things. From Example 2A.1 in Appendix 2A, the binary alphabet (square) minimum distance is 1.72, achieved with the error difference sequences $\Delta u_0, \Delta u_1 = 2, -2$ and $-2, 2$. Compared to orthogonal modulation, which has $d^2_{\min} = 2$, this represents an energy loss of $10\log_{10}(2/1.72) = 0.65\,\text{dB}$. With a 4-ary alphabet $d^2_{\min}$ is 0.69, orthogonal modulation achieves 0.8, and the loss is the same, 0.65 dB.

With the higher energies in narrowband signaling, error events leading to $d_{\min}$ dominate more and $Q(\sqrt{d^2_{\min} E_b/N_0})$ often provides an accurate enough estimate of symbol error rate. More accurate estimates are can be derived by taking into account that certain difference events are possible only with a few symbol sequences. *A priori* symbol probabilities can also be taken into account. The first issue is treated in Appendix 2A.

Model Phase Versions. An interesting fact is hidden in Eq. (2.45). As shown by Said (see Reference 10, or 8, Section 6.3), the minimization can be recast as

$$d^2_{\min} = \min_{\{\Delta u\}} (1/2E_b) \sum_n \sum_k \Delta u_k \rho_h[n-k] \Delta u_n, \qquad (2.48)$$

where ρ_h is the discrete-time autocorrelation of c. This implies that $d_{\min}$ and the whole distance structure of the signal set are *functions only of the autocorrelation of c*. There can be many c that lead to the same $\rho_h[k]$, and these are called *phase versions* of a model. They are easiest to describe in terms of $C(z)$, the z-transform of c. It is all-zero, with μ zeros in the z-domain; these occur in conjugate pairs (because $C(z)$ has real coefficients), and some are inside and some outside the unit circle. Consider a new c that has one or more zeros reflected about the unit circle, either from outside to inside or vice versa. If the old zero was located at z_o, $|z_o| \neq 1$, the

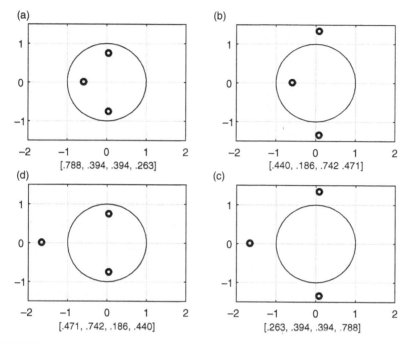

FIGURE 2.5   Zero locations for Example 2.2, showing all four-phase versions. (a) Min phase, (b) mid-phase version, (c) max phase, and (d) second mid phase. Discrete-time responses are shown under plots.

new one lies at $1/z_o$. It can be shown for $C(z)$ with real coefficients that the new model has the same autocorrelation. These reflection models are phase versions of each other. If $C(z)$ has v zeros that are not on the unit circle and not reflections or conjugates of another zero, then there are 2^v distinct phase versions.

The facts are made clear by the simple model just illustrated.

Example 2.2   *(Phase Versions)*
Let $c = [0.7878, 0.3939, 0.3939, 0.2629]$. Figure 2.5, a plots the zero locations of this $C(z)$; there is a conjugate pair and a negative real root, all inside the unit circle. There must be four-phase versions, since the pair can be inside or out and the real root can be inside or out. The figure shows all four, corresponding to the c taps [0.7878, 0.3939, 0.3939, 0.2629], [0.2629, 0.3939, 0.3939, 0.7878], [0.4397, 0.1860, 0.7421, 0.4706], and [0.4706, 0.7421, 0.1860, 0.4397].[7] Versions 1 and 2 and versions 3 and 4 are reverses of each other. The minimum distance of all versions is 1.72.

---

[7]Readers wishing to find phase versions for themselves can use the MATLAB function `roots` on the coefficients of $C(z)$, then reflect the roots, then use `poly` on the new roots to find the new phase version.

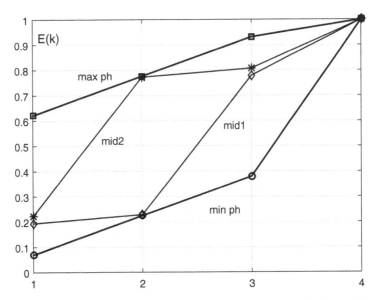

FIGURE 2.6 Partial energy plots for the four-phase versions in Figure 2.5.

Model phase has a major effect on decoding complexity, and this is the main reason to study it now. Phases are characterized as smaller or larger, with a smaller phase c being one that responds sooner, with more of its energy in the earlier taps. Reduced search decoders work by eliminating regions of the trellis, and this is most effectively done if the energy caused by a symbol comes earlier. A measure of this responsiveness is the *partial energy function*, defined by

$$E(k) = \sum_{\ell=0}^{k} |c_\ell|^2, \qquad k = 0, \dots, \mu. \qquad (2.49)$$

One is plotted in Figure 2.6 for each of the phase versions above. It can be seen that one underlies all the others; this is the minimum phase model, and it is the one with all zeros on or inside the unit circle. Another lies over all the others, has all noncircle zeros outside, and is the maximum phase model. Two other models, the "mid" phase models, lie in between these two.

As a rule a model of lesser phase may be converted to a model of greater phase through multiplication of its z-transform $C(z)$ by $F(z)$, where $F(z)$ is an allpass filter with poles at the (inside circle) positions to be reflected and zeros at the new positions. This can be done on noisy samples in the midst of detection because an allpass maintains the noise as white and a change of phase does not affect signal set distances. The reverse process, reducing the phase, is problematic because $F(z)$ will have poles outside the circle and be unstable. Actually there is no need to do this, because detecting the incoming signal backward reverses phase and converts maximum phase to the desired minimum phase. The receivers in later chapters that depend on minimum phase will convert to maximum phase and then detect the reverse signal.

If a model has zeros near but not on the unit circle, the different partial energy plots will appear in several tight groups, since reflecting the near zeros hardly changes $E(k)$. For example, if there are only three conjugate pairs inside and away from the circle, and the remaining zeros are close to the circle, there will be $2^3 = 8$ tight $E(k)$ groups. This phenomenon was studied by Balachandran [21], and plays a role in Said's optimal PRS in Chapter 7.

Spectral factorization is a method of solving for phase versions by splitting the sample autocorrelation function of c into two factors. The autocorrelation $\rho_c[-\mu], \ldots, \rho_c[0], \ldots, \rho_c[\mu]$ is symmetrical. Consequently, $R(z)$ has 2μ zeros that occur in conjugate reciprocal pairs z_o and $1/z_o^*$; zeros on the unit circle are double. It must be possible to factor $R(z)$ such that $R(z) = G(z)G^*(z^{-1})$, where $G(z)$ has one of each pair and and $G^*(z^{-1})$ has the other. If $z_1, \ldots, z_\mu$ are the roots of $G(z)$, then $1/z_1^*, \ldots, 1/z_\mu^*$ are the roots of $G^*(z^{-1})$. The coefficients of this $G(z)$ and $G^*(z^{-1})$ are two phase versions of c, one the reverse of the other.

Before leaving model phase we take on a more advanced model.

Example 2.3 *(Many Phase Versions)*
Consider a 30% root RC orthogonal pulse $f(t)$ that is scaled twice as long; that is, the new unit energy pulse is $h(t) = \sqrt{1/2}f(t/2)$. The spectrum of $h(t)$ is zero outside $1.3/4T$ Hz, T the symbol time. Pulses like this one will be important in Chapter 4. The receiver matched filter/sampler satisfies Property 2.1. The zeros[8] of the discrete-time response $h(nT)$, $T = 1$, are shown as circle markers in Figure 2.7. There are 4 zeros inside the circle and 4 reflections of these outside, plus one real zero inside and its reflection, and 8 zeros *on* the unit circle. The mid-phase response is plotted (circles) in the lower plot. If the circles strictly inside the unit circle are reflected outside, they join those already outside, and we have the zero plot of the maximum phase version; these zeros are marked with an "x." In this example, the new zeros land on old ones, giving 5 double zeros, marked with a "2," plus the original 8 on the unit circle. The corresponding maximum-phase response appears in the lower plot, marked with "x." It is clear that its energy arrives later than that of $h(nT)$.

OSB and WMF Models. For our purposes, a valid model is one that is coupled to white noise and leads to the distance structure of the original analog signals.

Different receivers lead in general to different discrete-time models of the same analog signal set. The model c implied by the OSB receiver is simply the modulator pulse samples $h(nT)$ given in Eq. (2.22), provided that OSB Property 2.1 holds. Even when it does not hold, Eq. (2.22) gives a valid model when $h(t)$ is constructed as

---

[8]The response has infinite duration. Response samples smaller than 0.006 are ignored, leaving 19 samples and an all-zero z-transform with 18 zeros. The five perfect reflections occur because $h(t)$ is symmetric in time.

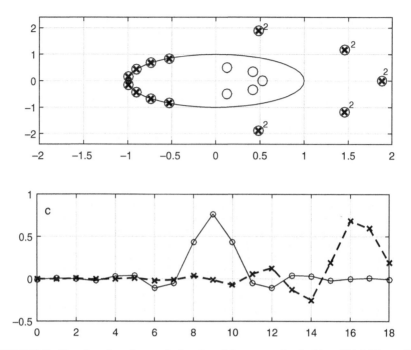

FIGURE 2.7   Zero locations (upper plot) and model response for double-width 30% root RC pulse (lower). "o" denotes original mid-phase pulse; "x" denotes max phase pulse version; "2" denotes a double zero at x. The dashed response below is max phase.

$h(t) = \sum c_\ell \varphi(t - \ell T)$, and $\{\varphi(t - \ell T)\}$ are an orthonormal signal space basis. In most of the sequel, we want to convert these models to maximum-phase models, using the procedures just discussed.

The matched filter receiver, Eq. (2.27), leads to its own length $2\mu + 1$ model $R_h(z)$, which is $\rho_h[k]$ as a time sequence, but this model is coupled to colored noise. With the WMF receiver, spectral factorization allows us to construct a white noise model of length $\mu + 1$. The required factorization of $R_h(z)$ is $G(z)G^*(z^{-1})$ in which the roots of $G(z)$ all lie outside the unit circle and those of $G^*(z^{-1})$ lie inside. The filter $G^*(z^{-1})$ then has a stable inverse and from Eq. (2.28), the coefficients of $G(z) = R_h(z)/G^*(z^{-1})$ form a time-discrete model. It is in fact maximum phase.

The fact that the model $G(z)$ has autocorrelation $R_h(z)$ is the key to proving $G(z)$ is optimal. All spectral factorizations of $R_h(z)$ lead to models $U(z)G(z)$ of the signal generation, including a $G(z)$ that might actually have been used at the transmitter. They and the actual continuous signals all create signal sets with the same distance structure; this can be verified by deriving the distance directly for a difference sequence Δu, which turns out in every case to be the same function of $R_h(z)$ only. Since the noise variates in the WMF receiver are AWGN with variance $N_0/2$, its statistical behavior must be the same as the ML receiver Eq. (2.6) that works with continuous signals. These facts are further demonstrated in the examples in Appendix 2A.

In applying the WMF procedure, we will require $R(z)$ to have no zeros on the unit circle, a condition that is generally true in applications when the pulse Fourier transform has no zeros in the range $[0, 1/2T)$ Hz. But the OSB model exists in principle under the opposite condition, when the transform bandwidth is less than $1/2T$ Hz. Thus the OSB and WMF models complement each other; generally speaking, each is valid when the other is invalid.

2.5   ERROR EVENTS

Codes and sophisticated modulations need to have memory from symbol to symbol, since otherwise coding gain and narrow bandwidth are impossible to achieve. Consequently, decoding and demodulation errors occur in multisymbol *events*. This section explores their properties. Much insight can be gained from them. The emphasis will be on events that occur in linear demodulation when the modulation pulse $h(t)$ is nonorthogonal and narrow band. Error events that occur in decoding trellis codes are similar, but these will be taken up as needed in later chapters.

2.5.1   Error Events and d_{min}

Error events and minimum distance for linear modulation were introduced earlier, together with the fact that demodulation error probability is $\approx Q(\sqrt{d_{min}^2 E_b/N_0})$. The event error rate (EER) is the rate of occurrence of whole events; it is driven primarily by d_{min} and the event multiplicity, which will be defined later in the section. The symbol error rate (SER) counts symbols, not events, and it is larger, since one symbol error occurs for each nonzero value in an event's difference sequence.

Two special error events are the one leading to d_{min} and the *antipodal* event. The second is the event with the difference sequence 2; that is, if u_1 is transmitted, the erroneous u_2 differs in only one place and by the least possible symbol difference. Antipodal events cause a single-symbol error. The antipodal square distance, denoted d_0^2, is numerically the same as the minimum distance of simple modulation, which was found in Eqs. (2.14)–(2.17) to be 2, 0.8 and 0.286 for binary, 4-ary, and 8-ary modulation. These values were found for orthogonal pulses, but the integrals are the same for any unit-energy pulse. The event pair in question is shown in Figure 2.8 for the nonorthogonal pulse in Example 2.2, $h(t) = \sqrt{1/2}f(t/2)$, $f(t)$ the 30% root RC pulse. For illustration, the transmitted symbols in the picture are $\ldots, 0, +1, 0, \ldots$ and $\ldots, 0, -1, 0, \ldots$, but the difference square energy is 2 for any pair of binary signals that differ in one place.

The following property holds.

Property 2.2   (Antipodal Signal Bound)   *For linear modulation with any pulse* $h(t)$, $d_{min}^2 \leq d_0^2$.

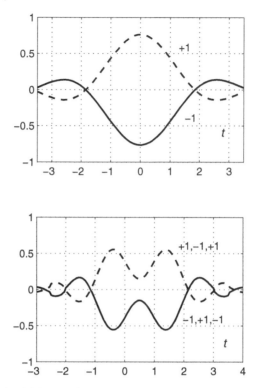

FIGURE 2.8   Antipodal (top) and $d_{\min}$-achieving events for a root RC pulse. Both pictures span seven symbol intervals. The antipodal difference [2] has square distance 2 and the $d_{\min}^2$ difference [2 -2 2] has 1.01.

The proof is simple: $d_{\min}$ cannot be greater than d_0 because it is the minimum of a set of events that include the antipodal one. $d_{\min}$ is often equal to d_0 even when the generator has $\mu \geq 1$; this is simply a statement that the symbol ISI can be removed by trellis detection. Unfortunately, significant band limitation normally leads to a smaller $d_{\min}$—if it did not, Shannon capacity would eventually be violated. For historical reasons, d_0^2 is often called the matched filter bound.[9] Whenever $h(t)$ is orthogonal on the symbol interval, $d_{\min}$ is d_0. Observe that $d_{\min}$ for a modulation has an upper bound, whereas $d_{\min}$ for a class of *codes* can be unbounded.

The minimum distance achieving difference needs to be found by a search over difference events. Routines to do this are presented in Appendix 2A. Figure 2.8 illustrates the minimum-achieving event for $h(t)$ in Example 2.3; the difference is [2 -2 2], which is illustrated by plotting the signals for symbols $\{\ldots, 0, +1, -1, +1, 0, \ldots\}$

---

[9]It was called so because in detection of simple modulation with orthogonal $h(t)$, the error probability $Q(\sqrt{d_0^2 E_b / N_0})$ is achieved only when the receiver filter is matched to $h(t)$.

and $\{\ldots,0,-1,+1,-1,0,\ldots\}$. $d_{\min}^2$ is 1.01 (see calculation in Example 2A.3). The difference [2 -2 2] holds for any phase version of $h(t)$.

Faster than Nyquist Pulses. In linear trellis modulation, a faster than Nyquist (FTN) pulse $h(t)$ is one that is T-orthogonal but is employed with a shorter symbol time τT, $\tau < 1$, for which it is not orthogonal; that is, the modulated signal is

$$s(t) = \sqrt{E_s} \sum u_n h(t - n\tau T), \qquad \tau < 1. \tag{2.50}$$

The PSD of $s(t)$ is the same with all τ (as in Eq. (1.7)). This is the accelerated time definition of FTN. An equivalent definition, the constant-interval definition, states that the symbol interval remains T and $h(t)$ is scaled longer by $1/\tau$:

$$s(t) = \sqrt{E_s} \sum u_n \sqrt{\tau} h(\tau t - nT). \tag{2.51}$$

This definition is sometimes more convenient.

Some history and properties of FTN signaling are given in Chapter 4. For now, the central fact about FTN is that its $d_{\min}^2$ continues to be the antipodal value d_0^2 for a considerable range of $\tau < 1$, this despite the fact that the per-databit bandwidth of the signal set declines by the factor τ. The explanation for how this can happen can be seen in the error difference events, specifically in the ones with the *second smallest* distance.

Take the $h(t)$ from Example 2.3, just discussed, as an illustration. This $h(t)$ is an FTN pulse with $\tau = 1/2$ and Figure 2.8 shows the difference event [2 -2 2] with distance 1.01. As τ is reduced from 1, an interesting phenomenon occurs. With $\tau = 1$ (orthogonal h), the square distance of the same difference event is 6 and the second smallest distance is 4, occurring with difference [2 -2], among others. With $\tau = 0.8$ the second smallest drops to 3.11 with difference [2 -2]; at $\tau = 0.703$ it is 2, the same as d_0^2, with difference [2 -2 2 -2 2 -2 2]. When τ reaches 0.5, it is the value shown in the figure, $d_{\min}^2 = 1.01$ with difference [2 -2 2]. In summary, the second smallest distance continually drops, and manifests itself with several difference events, but this has no effect on $d_{\min}$ until τ reaches 0.703. The gain in data throughput in the same bandwidth and the same approximate error rate is $1/0.703$, or 42%. This behavior––no change until a rather narrow bandwidth, then a progressive loss of distance––is typical of all $h(t)$, and roughly speaking, occurs even in nonlinear modulation. In Chapter 4, the critical τ is called the *Mazo limit*. It plays a role receiver complexity analysis as well.

As τ continues to drop, PSD bandwidth and $d_{\min}$ fall. To a degree, it is inevitable that these fall together since otherwise the Shannon limit will likely be exceeded. Figure 2.9 shows the relation between $d_{\min}^2$ and half-power bandwidth for the 30% root RC pulse FTN with 2-, 4-, and 8-ary modulation. The bandwidth in Hz-s/bit is $\tau / \log_2 M$, where M is the size of the modulation alphabet. The flat part of each curve ends at the Mazo limit and then drops off. An important observation is that M should change as the planned bandwidth is reduced: There are regions where 4-ary is better than binary modulation, and further to the left, regions where 8-ary is better

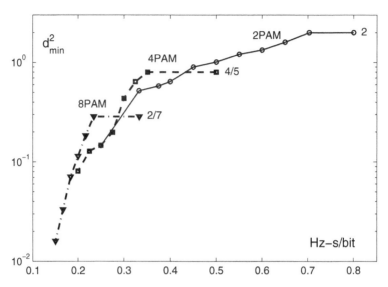

FIGURE 2.9 $d^2_{\min}$ versus half-power bandwdith for a 30% root RC FTN pulse and 2-, 4-, and 8-ary PAM modulation. (8PAM data adapted from Rusek[23].)

than 4-ary. The energy gain can be as much as 3 dB. That modulation needs to adjust to the bandwidth requirement is a theme throughout the book.

Despite the word Nyquist, the foregoing discussion actually has nothing to do with orthogonality. *All* pulses in linear modulation have distance d^2_0 with difference event 2. If initially $d_{\min}$ is d_0, and we scale $h(t)$ longer, the pulse will eventually reach its Mazo limit and $d_{\min}$ will drop below d_0. The FTN concept—$d_{\min}$ equal d_0, then a drop—thus applies to many nonorthogonal pulses. In Chapter 7, we will see cases where the most desirable pulses are not T-orthogonal for any T. The case where there is initial T-orthogonality will be called *classical* FTN.

Error Event Multiplicity. The set of distances larger than $d_{\min}$ and their difference events play another important role. It is possible to sharpen the error probability estimate $Q(\sqrt{d^2_{\min}E_b/N_0})$ by adding more terms for less common events. The result will be a union bound estimate, and is therefore an upper bound, but E_b/N_0 is higher with narrowband signals and this lessens the error in the bound. Another sharpening of the bound takes place because of the *multiplicity* parameter κ of an error event. This measures the input symbol sequences for which an error difference can possibly occur. By definition κ is the number of nonzero elements in a difference, minus 1.

The effect of κ is best explained by an example. Taking the case in Figure 2.8, we see that if symbols $\boldsymbol{u}_o = 1\ 1\ 1\ \dots$ were modulated and $\Delta\boldsymbol{u} = \boldsymbol{u}_o - \boldsymbol{u}_e = [2\ \text{-}2\ 2]$, $\boldsymbol{u}_e$ an erroneous sequence, there is no $\boldsymbol{u}_e$ that leads to $[2\ \text{-}2\ 2]$. In fact, only the modulation symbols $1\ \text{-}1\ 1$ allow this. That is, only 1/4 of transmitted symbols, namely $2^{-\kappa}$, can lead to difference $[2\ \text{-}2\ 2]$. If the difference were $[2\ 0\ 2]$, only $\boldsymbol{u}_e = 1\ \text{X}\ \text{-}1\ \dots$ would be forbidden, where X is the same symbol to the one in $\boldsymbol{u}_o$. This is half the sequences,

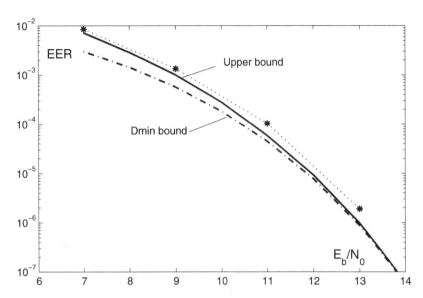

FIGURE 2.10 Comparison of two EER estimates to actual performances (*). Demodulation with $\tau = 1/2$ FTN pulse response.

or again $2^{-\kappa}$. We do not consider the first symbol because the linearity of the signal generation means that it may always be taken as $+1$.

Precisely speaking, the error probability calculated here is defined to be the probability that a decoder seeking the closest signal to $r(t)$ suffers an error event starting at symbol $n_o + 1$, given that detection is correct up to n_o. The expression $Q(\sqrt{d^2 E_b/N_0})/2^{\kappa}$ measures the probability of a certain difference event $\mathcal{E}$ that has square distance d^2. The rate of all such events is the event error rate, or EER. If $\mathcal{E}$ is a $d_{\min}$-achieving event, we have an underbound; if a sum is formed over a few more events, we have a larger, usually better estimate but it is not necessarily an overbound; a sum over all significant events yields an overbound. In receiver tests, the EER is measured by counting events and dividing by the opportunities for a new event to start; that is, symbol times during ongoing events are subtracted from the symbol total. Attending to this detail allows a better match to theory at high EERs, although it matters little at small EERs.

If $\mathcal{E}$ occurs, there will be modulation symbol errors at its nonzero difference positions. The expectation is $(\kappa + 1)Q(\sqrt{d_{\min}^2 E_b/N_0})$ errors. By summing such contributions, we obtain a lower bound, then an estimate, in the manner above for the symbol error rate (SER). The routine `mlsedist2Q` in Appendix 2A computes such EER and SER bounds at a given E_b/N_0. Figure 2.10 compares the EER bounds to the actual event rate of a near-optimal BCJR demodulator from Chapter 4 when the pulse $h(t)$ is the nonorthogonal one in Example 2.3. The estimate is constructed from the union of all events with $d^2 < 2$.

2.5.2   Error Fourier Spectra

Until now we have computed distances by integrating over time, but the calculation can just as well be performed over frequency. The outcome gives insight into where distance lies in the transmission bandwidth, and in particular, into how band limitation will constrain distance.[10]

Let D be the non-normalized distance of an error event. From Parseval's identity,

$$D^2(s_1(t), s_2(t)) = \int |s_1(t) - s_2(t)|^2 dt = \int |S_1(f) - S_2(f)|^2 df. \qquad (2.52)$$

The last difference we can write as $\Delta S(f) = S_1(f) - S_2(f)$. The quantity $|\Delta S(f)|^2$ is the *difference power spectrum* of the event.

The technique can be carried another step. Suppose a subsequent linear filter with impulse response $g(t)$ limits the signal bandwidth further. Then the difference event square distance simply converts to

$$\int |s_1(t) \cdot g(t) - s_2(t) \cdot g(t)|^2 dt = \int |S_1(f) - S_2(f)|^2 |G(f)|^2 df$$

$$= \int |\Delta S(f)|^2 |G(f)|^2 df. \qquad (2.53)$$

The old difference spectrum is just filtered to produce the new spectrum. By checking Eq. (2.53), or by computing d_{min} for the new pulse response $g \cdot h$ (by a program in Appendix 2A), it will be clear if the filter is seriously reducing error performance. If $G(f)$ is an ideal low-pass filter with passband $(-W, W)$, the last integral is simply $\int_{-W}^{W} |\Delta S(f)|^2 df$, and we have the distribution of distance as a function of frequency.

The convolution with $g(t)$ can also be viewed as creating a new nonorthogonal modulation. All of these insights will be applied in later chapters.

Example 2.4   *(Escaped Distance with a Narrowband Pulse)*

When a pulse is very narrowband, there can be dramatic effects at the modulation band edge. Figure 2.11 shows the difference power spectra of two error events when $h(t) = \sqrt{1/4}\, f(t/4)$ and $f(t)$ is the 30% root RC pulse orthogonal on $T = 1$. Since the picture shows positive frequencies only, the integral of the power spectra is half their respective distances; the events shown are the antipodal event, square distance 2, and the d_{min} event [2 -2 -2 2 2 -2] with distance 0.218. A vertical line marks 0.125 Hz, the half-power bandwidth of the modulation. Were the signal spectrum limited to 0.125 Hz, it would hardly affect the antipodal event, but it would remove 54% of the d_{min} event's square distance, a loss equivalent to 3.3 dB in E_b/N_0.

What happens in the preceeding example, whereby a significant part of an event's distance pushes outside what would seem to be the signal band edge, is called *escaped distance*. Severe instances can occur, and will be noted in later chapters. They mean

---

[10]The ideas in Section 2.5.2 developed in the 1980s from the work of the author and N. Seshadri [24,25].

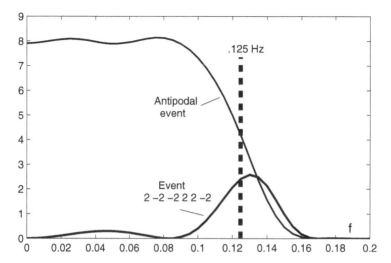

FIGURE 2.11 Antipodal and $d_{\min}$-achieving difference event spectra with $\tau = 1/4$ FTN pulse modulation. $d_{\min}$ event is [2 -2 -2 2 2 -2]. The PSD of the modulation itself is 1/2 that of the antipodal event.

that the difference between signals lies at a much wider bandwidth than the main body of the signals. Often, but not always, the critical differences are the ones that lead to $d_{\min}$, and if much of their distance is truly lost from the receiver's view, the effective $d_{\min}$ is much degraded.

A criterion for such a critical difference is clear in Figure 2.11: Low spectrum at low frequencies, or more specifically, a DC null in the difference spectrum. If a null exists, distance must necessarily be forced outward in frequency; if the signaling is very narrowband, $d_{\min}$ is very small and can in principle lie in the stop band of the signal spectrum. A straightforward Fourier analysis shows that a DC null is equivalent to $\int \sum \Delta u_n h(t - nT) dt = 0$. A sufficient condition for this is the zero sum rule

$$\sum_n \Delta u_n = 0. \tag{2.54}$$

Nulls can also originate from an oddly shaped $h(t)$, but in fact Eq. (2.54) holds almost always for the minimum distance when bandwidth is narrow. Both the DC null and the zero sum hold in Figure 2.11 and the zero sum rule holds in fact for almost all near contenders for the minimum distance in Example 2.4. By limiting the search for $d_{\min}$ to zero-sum differences, the search can be much reduced.

A summary of the $d_{\min}$ situation is as follows. If the modulation pulse $h(t)$ is orthogonal in the symbol time, the minimum distance is the antipodal d_0 and all other difference events have square distance a multiple of d_0^2. As $h(t)$ scales longer, and declines presumably in bandwidth, $d_{\min} = d_0$ until the Mazo limit for the pulse is crossed. Thereafter, $d_{\min}$ drops and eventually is the distance of a zero-sum difference. Very roughly, this begins to occur at about one third of the orthogonal pulse bandwidth. A number of bounds to d_{min} based on these spectral ideas are to be found in Reference 8, Chapter 6 and in Reference 10, 22, and 25.

2.6 CONCLUSIONS

Chapter 2 began with signal space theory for the white Gaussian noise channel. This important theory expresses signals as components in a vector space and error probability in terms of Euclidean distance. The remainder of the book is based on this approach. The theory leads to a number of optimal receiver structures, the most important of which in bandwidth efficient coding is the orthogonal simple basis (OSB) receiver.

The OSB receiver expresses interval-T linear modulation signals as weighted sums of simple T-orthogonal basis signals. The linear modulation pulse is not necessarily T-orthogonal but the basis functions are. The receiver begins with a simple matched filter, matched to the basis function, and its T-samples are noise plus the real-number convolution of a generator sequence and the data. The rest of the receiver is an all-pass filter, which produces a maximum-phase version of the samples, and a trellis decoder.

Because strong bandwidth limitation causes strong Markov intersymbol interference, a trellis decoder is essential in bandwidth efficient signaling. Furthermore, most coded bandwidth efficient signals today need an iterative decoder. This consists of two soft-output trellis decoders that feed likelihoods to each other. The BCJR soft trellis decoder is an important such decoder.

The most important narrowband coding schemes today are in the faster than Nyquist class. These are based on nonorthogonal linear modulation, combined with convolutional coding. One BCJR algorithm decodes the convolutional codewords while a second BCJR demodulates the ISI. The first BCJR is traditional and simple but the second must be simplified; a number of methods are available.

The phase of a received sequence is an important concept in simplified iterative decoding. Receivers with reasonable complexity require maximum phase sequences, which they then reverse to obtain minimum phase. Spectral aliasing is another important concept in signal design.

Trellis detection produces multiple-symbol error events instead of independent data errors. These have a minimum distance, and other properties such as multiplicity and difference spectrum, that are useful in fine-tuning error rate predictions.

APPENDIX 2A: CALCULATING MINIMUM DISTANCE

An important tool in the analysis of codes and modulations is the calculation of minimum distance. This appendix presents Matlab scripts for several algorithms that find it for trellis signal structures. Examples are given that illustrate the Mazo limit and when OBD and WMF receivers are not optimal.

Program 2A.1: Minimum Distance of Binary Linear Modulation

The next MATLAB script finds the (square) minimum distance for binary baseband linear modulation with a basis pulse $h(t)$. It works by computing the distance for each symbol difference sequence $\Delta\mathbf{u}$ of length up to totsym and noting the minimum. The last difference at each length must be nonzero and the first is fixed at 2 since the distance of $\Delta\mathbf{u}$ and $-\Delta\mathbf{u}$ are the same. Technically, this enumeration does not prove that the minimum distance is the value found, but distances grow rapidly with difference length, and with modern computing this limited exhaustive search is a clean, fast way to estimate the minimum. Algorithms that find a guaranteed minimum distance are available in the literature, based for example on a trellis search.

The program can work with either continuous or discrete-time h, given in both cases in the input hh. Which is the case is specified in the sampling rate fs; setting fs=1 means h is discrete-time. hh is the same sequence whether or not $h(t)$ is causal, but the routine may reach a conclusion faster if a minimum phase h is given. If the signals are filtered by a filter with impulse response gg, the input pulse response is then hh=conv(hh,gg). If continuous, hh should begin and end at a symbol interval boundary.

The output is the minimum distance plus all distances found less than distlim and their difference sequences.

```
%   mlsedist2 is a script to compute the minimum square free distance of
%   binary PAM modulation with total pulse response 'hh'. It computes
%   the distance for each symbol difference sequence up to length 'totsym'
%   and finds the min of these. Supply the following:
%     fs = sampling rate per symbol interval (set to 1 if discrete time)
%     hh = total pulse response or discrete time model
%     totsym = length limit of difference sequences
%     distlim = limit on distance of print out

dsmap=[-2 0 2]; mindist=2; mindel=2; z=.5/fs;          %Initialize
hh=hh/sqrt(sum(hh.^2)/fs);                             %Normalize
dist=2;                                                %Antipodal distance
%       Do short sequences first. Least square distance so far is in mindist.
disp(['Error seqs and distances found inside distlim ...'])
disp(['Antipodal signal distance is 2']); df=zeros(1,fs+1);
delta=[2-2]; df(1:fs:length(df))=delta; dist=z*sum(conv(hh,df).^2);
  if dist < mindist, mindist=dist; mindel=delta;  end
  if dist < distlim, disp(dist), disp(delta),  end
delta=[2 2]; df(1:fs:length(df))=delta; dist=z*sum(conv(hh,df).^2);
  if dist < mindist, mindist=dist; mindel=delta;  end
  if dist < distlim, disp(dist), disp(delta),  end

%       Do long sequences. Enumerate middle parts of lengths kk = 1,...,totsym-2.
for kk=1:totsym-2,
  pwr(1:kk)=3.^(kk-1:-1:0); sym=zeros(1,kk);
  df=zeros(1,(kk+1)*fs+1); ldf=length(df);
  for nct=0:3^kk-1, tm=nct;                            %Enumerate middle sections
    for k=1:kk,                                        %Find difference symbols
      midk=floor(tm/pwr(k));
      tm=tm-midk*pwr(k);
      sym(k)=dsmap(midk+1);  end
%       Append beginning and end symbols and check distance
    delta=[2 sym -2]; df(1:fs:ldf)=delta; dist=z*sum(conv(hh,df).^2);
      if dist < mindist, mindist=dist; mindel=delta;  end
      if dist < distlim, disp(dist), disp(delta),  end
```

```
    delta=[2 sym 2];  df(1:fs:ldf)=delta; dist=z*sum(conv(hh,df).^2);
      if dist < mindist, mindist=dist; mindel=delta;  end
      if dist < distlim, disp(dist), disp(delta),  end
end, end
%                         Final output
disp('Minimum square distance found and delta are ...')
disp(mindist), disp(mindel), distlim, totsym
```

Example 2A.1 *(Example 2.2 Continued)*

The discrete-time pulse is hh = [.7878, .3939, .3939, .2629]. With distlim=2.5; totsym=6; fs=1, mlsedist2 gives d^2_{min} = 1.72, achieved with error difference sequence [2 -2]. It also finds another sequence [2 0 -2] with distance 2.34, and no others with distance below 2.5. Large values of totsym do not change the outcome. All phase versions give the same distribution of outcomes.

Example 2A.2 *(Example 2.2 with Continuous Signals)*

Now suppose $h(t) = .7878f(t) + .3939f(t-1) + .3939f(t-2) + .2629f(t-3)$, where $f(t)$ is the $T = 1$ orthogonal 30% root RC pulse. This is the previous discrete-time response convolved with the root RC pulse. mlsedist2 with fs=20 and the sampled new pulse yields the same outcome as before. This happens because $f(t)$ is a weighted sum of orthogonal pulses.

Example 2A.3 *(Non-Orthogonal Pulse)*

For a third example let $h(t) = \sqrt{1/2}f(t/2)$ from Example 2.3, where $f(t)$ is the root RC pulse in Example 2.2 scaled twice as wide in time. Take mlsedist2 with distlim=1.5; totsym=8; fs=20 yields minimum distance 1.01, achieved by difference [2 -2 2]. Eight more differences lead to distances less than 1.5. Exactly the same outcome occurs when the continuous $h(t)$ is replaced by 15 of its samples each $T = 1$ seconds, the middle seven of which are [-.109 -.053 .435 .765 .435 -.053 -.109]. This happens because $h(t)$ and a simple basis set satisfy Property 2.1.

Example 2A.4 *(Pulse that Fails Property 2.1)*

This example has three parts. (*i*) Change the pulse in the previous example to $h(t) = \sqrt{2/3}\,f(2t/3)$, where $f(t)$ now has excess bandwidth factor 0.7. This pulse has bandwidth 13% larger than the Nyquist bandwidth $1/2T$ Hz, and fails Property 2.1 for any simple orthonormal basis. Strictly speaking, only the WMF receiver can be used. The continuous autocorrelation $\rho_h(\delta)$ is $\int h(x)h(x+\delta)\,dt$ and we work with its samples $\rho_h(kT)$, k an integer, $T = 1$. Taking the central 29 of these, we find with roots that there are 28 roots, 14 outside the unit circle, 14 reflected inside, and none on the circle (the reflections are because $\rho_h(kT)$ is symmetric in k). The 14 outside the circle are the roots of the discrete-time maximum phase model $G(z)$, whose coefficients (found with poly and reversed) turn out to be [.906 .413 -.093 .0046 -.0094 .0026 ...]. Applying this in mlsedist2 with distlim=3; totsym=8; fs=1, we get that that d^2_{min} = 2, due to the antipodal difference event, with difference

[2 -2] achieving 2.66 and [2 -2 2] achieving 2.99. Note that it is $\rho_h(\delta)$ that is sampled in the WMF modeling, not $h(t)$.

(*ii*) For comparison, we can compute the OSB model, which is based on the samples $h(kT)$. Taking the middle 15 of these samples, finding again 14 roots (none on the unit circle), reflecting those inside to the outside, and applying `poly`, we compute a somewhat different model [.9265 .370 -.067 -.0155 .0076 -.0107 ...]. It also leads to $d_{min}^2 = 2$, but the next-nearest difference is at distance 2.73 and a rather different distribution exists thereafter. The OSB model is not valid here, although the d_{min}^2 outcome is the same.

(*iii*) How can we verify that the first calculation is valid? By running `mlsedist2` with the original continuous $h(t)$, this is done. Setting `distlim=3; totsym=8; fs=20`, one gets exactly the result in (*i*).

Program 2A.2: Minimum Distance of 4-Level Linear Modulation

The next MATLAB script is similar to the previous one, but extends the method to 4-ary baseband modulation. Since the modulator symbol set is now $\sqrt{1/5}\,\{3, 1, -1, -3\}$, the symbol differences to take account of are $\sqrt{1/5}\,\{0, \pm2, \pm4, \pm6\}$.

```
%    mlsedist4 is a script to compute the minimum square free distance of
% 4PAM modulation with total pulse response 'hh'. It computes the
% distance for each symbol difference sequence up to length 'totsym'
% and finds the min of these. Supply the following:
%    fs = sampling rate per interval (set to 1 if discrete time)
%    hh = total impulse response or discrete time model
%    totsym = length limit of difference sequences
%    distlim = limit on distance of print out

   dsmap=[-6 -4 -2 0 2 4 6]; mindist=.8; mindel=2; z=.2/fs;  %Initialize
   hh=hh/sqrt(sum(hh.^2)/fs);                          %Normalize
   dist=.8;                                            %Antipodal distance
   ends2=[2 -6;2 -4;2 -2;2 2;2 4;2 6;4 -6;4 -2;4 2;4 6;6 -4;6 -2;6 2;6 4];
   ends=[ends2;  4 -4;4 4;6 -6;6 6];

%      Do short sequences first. Least square distance so far is in mindist.
%      There are 14 2-symbol differences (4 more give larger distances).
 disp(['Error seqs and distances found inside distlim ...'])
 disp(['Antipodal signal distance is .8'])
 df=zeros(1,fs+1);
for n=1:14,  del=ends2(n,:);
  df(1:fs:length(df))=del; dist=z*sum(conv(hh,df).^2);
  if dist < mindist, mindist=dist; mindel=del;   end
  if dist < distlim, disp(dist), disp(del),   end
end

%      Do long sequences. Enumerate middle parts of lengths kk = 1,...,totsym-2.
%      There are 18 end-symbol pairs for each middle part.
for kk=1:totsym-2,  k1=kk+2;
  pwr(1:kk)=7.^(kk-1:-1:0); sym=zeros(1,kk);
  df=zeros(1,(kk+1)*fs+1); ldf=length(df);
  for nct=0:7^kk-1,                                    %Eumerate middle sections
    for k=1:kk, tm=nct;                                %find difference symbols
      midk=floor(tm/pwr(k));
      tm=tm-midk*pwr(k);
      sym(k)=dsmap(midk+1);   end
```

```
%         Append beginning and end symbols and find distances.
   for n=1:18,   del=[ends(n,1) sym ends(n,2)];
      df(1:fs:ldf)=del; dist=z*sum(conv(hh,df).^2);
      if dist < mindist, mindist=dist; mindel=del;   end
      if dist < distlim, disp(dist), disp(del),   end
end, end, end
%                        Print final output
disp('Minimum square distance found and delta are ...')
disp(mindist), disp(mindel), distlim, totsym
```

Example 2.1 may be repeated for 4-ary modulation with $\mathtt{mlsedist4}$ in place of $\mathtt{mlsedist2}$, with the outcome that $d^2_{\min} = .69$ with difference event [2 -2]. This is the same 1.72 as in Example 2.1, but scaled by $.8/2$ to reflect the new symbol energy.

Repeating Example 2.3 with $\mathtt{distlim=.5}$; $\mathtt{totsym=10}$; $\mathtt{fs=20}$ and the continuous $h(t)$ yields $d^2_{\min} = .154$ for difference sequence [2 -4 6 -6 4 -2]. The discrete-time h there yields the same with $\mathtt{fs=1}$. The difference [2 -2 2] has distance .404, which is again the energy-scaled distance 1.01 that was found before. But this time the 4-ary minimum distance is lower than would be predicted from the binary case, and it occurs with a difference that definitely stems from 4-ary symbol sequences.

Program 2A.3: BER and EER Estimates for Binary Linear Modulation

This Matlab script extends the method in Program 2A.1 to computing the event error rate (EER) and symbol bit error rate (BER) for binary modulation, using the multiplicity parameter κ defined in Section 2.5.1 and the Gaussian integral $Q(\ )$. Error difference events leading to square distances less than $\mathtt{distlim}$ count in the estimates. Each creates a term of the form $Q(\sqrt{d^2 E_b/N_0})/2^\kappa$; this alone counts in the EER sum, and the BER sum includes a factor to account for the event's symbol errors. The E_b/N_0 for the Q calculation is specified in $\mathtt{snr}$. Otherwise, the inputs are as in Program 2A.1, with the input $\mathtt{fs}$ specifying whether the pulse response is continuous or discrete.

EER and BER will be based on the $d_{\min}$-achieving event alone if $\mathtt{distlim}$ is set slightly above $d^2_{\min}$. The program employs the function $\mathtt{qfn(x)}$ to evaluate $Q(x)$. If this is not available, use $\mathtt{qfn(x)=.5-.5*erf(x/1.4142136)}$.

```
% mlsedist2Q is a script to compute for binary PAM modulation with total
% pulse response 'hh' the minimum square free distance, event error rate (EER)
% and bit error rate (BER) based on event multiplicity. It computes the
% distance for each symbol difference sequence up to length 'totsym', finds
% the min of these, and constructs estimates based on Q( ) for distances less
% than 'distlim'.  Supply the following:
%    fs = sampling rate per symbol interval (set to 1 if discrete time)
%    hh = total pulse response or discrete time model
%    totsym = length limit of difference sequence
%    distlim = limit on distance of print out and Q calculation
%    snr = Eb/No for Q calculation (not in dB)

   dsmap=[-2 0 2]; dist=2; mindist=2; mindel=2; z=.5/fs;      %Initialize
   hh=hh/sqrt(sum(hh.^2)/fs);                                  %Normalize
```

```
%     Do short sequences first. Least square distance so far is in mindist.
%     For each, find Q term, divide by multiplicity, find EER & BER term, sum.
disp(['Q contributions inside distlim found at SNR ',num2str(snr)])
disp(['Distances, diff. seqs., Q term, EER & BER terms, total EER & BER'])
if distlim > 2,                                      %Antipodal case
  qtot=qfn(sqrt(snr*2)); btot=qtot;
    disp([' Distance 2, d-seq 2,   ']), disp([qtot btot qtot btot])
else qtot=0; btot=0;  end
df=zeros(1,fs+1); delta=[2 -2];
df(1:fs:length(df))=delta; dist=z*sum(conv(hh,df).^2);
  if dist < mindist, mindist=dist; mindel=delta;  end
  if dist < distlim, tq=qfn(sqrt(dist*snr))/2;
    qtot=qtot+tq; tb=2*tq; btot=btot+tb;
    disp([' Distance ',num2str(dist),',  d-seq ',int2str(delta)])
    disp(['     ',num2str([tq tb qtot btot])]),  end
delta=[2 2]; df(1:fs:length(df))=delta; dist=z*sum(conv(hh,df).^2);
  if dist < mindist, mindist=dist; mindel=delta;  end
  if dist < distlim, tq=qfn(sqrt(dist*snr))/2;
    qtot=qtot+tq; tb=2*tq; btot=btot+tb;
    disp([' Distance ',num2str(dist),',  d-seq ',int2str(delta)])
    disp(['     ',num2str([tq tb qtot btot])]),  end

%     Do long sequences. Enumerate middle parts of length kk = 1,...,totsym-2.
for kk=1:totsym-2,
  pwr(1:kk)=3.^(kk-1:-1:0); sym=zeros(1,kk);
  df=zeros(1,(kk+1)*fs+1); ldf=length(df);
  for nct=0:3^kk-1, tm=nct;                          %Enumerate middle sections
    for k=1:kk,                                      %Find difference symbols
      midk=floor(tm/pwr(k));
      tm=tm-midk*pwr(k);
      sym(k)=dsmap(midk+1);  end
%         Append beginning and end symbols; find multiplicity 'mlt' and distance
      delta=[2 sym -2]; df(1:fs:ldf)=delta; dist=z*sum(conv(hh,df).^2);
        if dist < mindist, mindist=dist; mindel=delta;  end
        if dist < distlim, mlt=sum(delta~=0)-1;
          tq=qfn(sqrt(dist*snr))/(2^mlt);
          qtot=qtot+tq; tb=(mlt+1)*tq; btot=btot+tb;
          disp([' Distance ',num2str(dist),',  d-seq ',int2str(delta)])
          disp(['     ',num2str([tq tb qtot btot])]),  end
      delta=[2 sym 2]; df(1:fs:ldf)=delta; dist=z*sum(conv(hh,df).^2);
        if dist < mindist, mindist=dist; mindel=delta;  end
        if dist < distlim, mlt=sum(delta~=0)-1;
          tq=qfn(sqrt(dist*snr))/(2^mlt);
          qtot=qtot+tq; tb=(mlt+1)*tq; btot=btot+tb;
          disp([' Distance ',num2str(dist),',  d-seq ',int2str(delta)])
          disp(['     ',num2str([tq tb qtot btot])]),  end
end, end
%                     Final output
disp('Minimum square distance found & difference are'),
  disp(['  ',num2str(mindist),' at ',int2str(mindel)])
disp('EER & BER estimates at SNR are'), disp([qtot btot])
disp(['Distance limit = ',num2str(distlim)])
disp(['Symbols searched = ',num2str(totsym)])
```

The program above can be adopted to 4-ary modulation in the style of Program 2A.2, and we leave this as an exercise for the reader.

REFERENCES

1. *J.B. Anderson, *Digital Transmission Engineering*, 2nd ed., Wiley–IEEE Press, Piscataway, NJ, 2005.

2. *J.G. Proakis, *Digital Communication*, 4th and later eds., McGraw-Hill, New York, 1995.

3. *J.M. Wozencraft and I.M. Jacobs, *Principles of Communication Engineering*, Wiley, New York, 1965.

4. *M. Schwartz, *Information Transmission, Modulation, and Noise*, 4th ed., McGraw-Hill, New York, 1990.

5. V.A. Kotelnikov, The theory of optimum noise immunity, Ph.D. thesis, Molotov Energy Institute, Moscow, 1947; available under same title from Dover Books, New York, 1968 (R.A. Silverman, translator).

6. C.E. Shannon, Communication in the presence of noise, *Proc. IRE*, **37**, pp. 10–21, 1949; reprinted in *Claude Elwood Shannon: Collected Papers*, Sloane and Wyner, eds., IEEE Press, New York, 1993.

7. A. Papoulis, *Probability, Random Variables and Stochastic Processes*, 3rd ed., McGraw-Hill, New York, 1991.

8. J.B. Anderson, A. Svensson, *Coded Modulation Systems*, Kluwer-Plenum, New York, 2003.

9. G.D. Forney, Jr., Maximum-likelihood sequence estimation of digital sequences in the presence of intersymbol interference, *IEEE Trans. Inf. Theory*, **18**, pp. 363–378, 1972.

10. A. Said, Design of optimal signals for bandwidth efficient linear coded modulation, Ph.D. thesis, Dept. Electrical, Computer and Systems Eng., Rensselaer Poly. Inst., Troy, NY, 1994.

11. B. Hassibi and H. Vikalo, On the sphere-decoding algorithm I. Expected complexity, *IEEE Trans. Signal Proc.*, **53**, pp. 2806–2818, 2005.

12. J. Jalden and B. Ottersten, On the complexity of sphere decoding in digital communications, *IEEE Trans. Signal Proc.*, **53**, pp. 1474–484, 2005.

13. P. Robertson, E. Villebrun, and P. Hoeher, A comparison of optimal and sub-optimal map decoding algorithms operating in the log domain, *Conf. Record*, IEEE Int. Conf. Communications, Seattle, pp. 1009–1013, 1995.

14. I. Kanaras, A. Chorti, et al., Spectrally efficient OFDM signals: Bandwidth gain at the expense of receiver complexity, in *Proc. IEEE Int. Conf. Communications*, Dresden, 2009.

15. L.R. Bahl, J. Cocke, F. Jelinek, and J. Raviv, Optimal decoding of linear codes for minimizing symbol error rate, *IEEE Trans. Information Theory*, vol. IT-20, pp. 284–287, 1974.

16. R. Koetter, A.C. Singer, and M. Tüchler, Turbo equalization, *IEEE Signal Proc. Mag.*, **21**, pp. 67–80, 2004.

17. J.B. Anderson and S.M. Hladik, Tailbiting MAP decoders, *IEEE J. Sel. Areas Commun.*, **16**, pp. 297–302, 1998.

---

*References marked with an asterisk are recommended as supplementary reading.

18. J.B. Anderson and K.E. Tepe, Properties of the tailbiting BCJR decoder, in *Codes, Systems, and Graphical Models*, B. Marcus and J. Rosenthal, eds, Springer, New York, pp. 211–238, 1999.

19. G. Colavolpe and A. Barbieri, On MAP symbol detection for ISI channels using the Ungerboeck observation model, *IEEE Commun. Lett.*, **9**, pp. 720–722, 2005.

20. R.A. Gibby and J.W. Smith, Some extensions of Nyquist's telegraph transmission theory, *Bell Sys. Tech. J.*, **44**, pp. 1487–1510, 1965.

21. K. Balachandran and J.B. Anderson, Reduced complexity sequence detection for nonminimum phase intersymbol interference channels, *IEEE Trans. Inf. Theory*, **43**, pp.275–280, 1997.

22. A. Said and J.B. Anderson, Bandwidth efficient coded modulation with optimized linear partial-response signals, *IEEE Trans. Inf. Theory*, **44**, pp. 701–713, 1998.

23. F. Rusek, Partial response and faster-than-Nyquist signaling, Ph.D. thesis, Electrical and Information Technology Dept., Lund University, Lund, Sweden, 2007.

24. N. Seshadri, Error performance of trellis modulation codes on channels with severe intersymbol interference, Ph.D. thesis, Dept. Electrical, Computer and System Eng., Rensselaer Poly. Inst., Troy, N.Y., 1986.

25. N. Seshadri and J.B. Anderson, Asymptotic error performance of modulation codes in the presence of severe intersymbol interference, *IEEE Trans. Inf. Theory*, **34**, pp. 1203–1216, 1988.

3

GAUSSIAN CHANNEL CAPACITY

INTRODUCTION

A central outcome of Shannon information theory is limits to communication in the form of capacity formulas. These can be written as a function of the channel structure and the transmission symbols employed. Our purpose in this chapter is as it was in the last, to outline basic ideas, emphasizing those that are important in narrowband transmission over the Gaussian noise channel.

We need to redirect the classical theory in several ways. Bandwidth plays a critical role and needs explicit expression, not in bits per second but in bits per Hertz-second. Not only the width but also the shape of the signal PSD affect the outcome, especially when the transmission employs nonorthogonal pulses. In the later chapters, we will see practical systems that touch the traditional Shannon limit when they should not, and researchers have found systems the even "exceed" limits. The problem is usually a weak definition of bandwidth.

Second, we would like to express the Shannon limits as data bit error rate (BER) obtainable with the data E_b/N_0, since this is the measure of real codes and modulations.

Section 3.1 reviews the standard approach to channel capacity, ending with the capacity for the single-use AWGN channel. Section 3.2 extends that to a continuous-signal channel with a bandwidth measure, an arbitrary PSD, and a specified data BER. This creates a Shannon limit that directly compares to concrete transmission systems in the next chapters. Section 3.3 looks at important implications for narrowband

Bandwidth Efficient Coding, First Edition. John B. Anderson.
© 2017 by The Institute of Electrical and Electronics Engineers, Inc. Published 2017 by John Wiley & Sons, Inc.

systems. One result is that most often orthogonal pulse transmission has a weaker capacity for the same spectrum than nonorthogonal.

Much of the needed Shannon theory is only summarized here, and many good text books provide details. Classic texts are Gallager [1] and Cover–Thomas [2]; the more recent Yeung [3] emphasizes the Gaussian channel. The communication engineering texts [4,5,7] have useful sections, and Wilson [7] is especially clear about the effect of alphabet constraints.

3.1   CLASSICAL CHANNEL CAPACITY

We begin with the channel capacity of a single use of an AWGN channel. In the language of Chapter 2, a value x is transmitted, a zero-mean IID Gaussian variate η with variance $N_0/2$ is added, and the received value is $r = x + \eta$. This constitutes one *channel use*. The variable x has average energy E_{cu}, joules per channel use. The rest of this section calculates the capacity of x and then finds the number of such uses available in a piece of time and bandwidth. The product is the capacity of the piece.

In Shannon theory the formal definition of capacity is

$$C = \max_{P[X]} I(X; R) \quad \text{bits/channel use,} \tag{3.1}$$

where

$$I(X; R) = \sum_x \sum_y P[x, r] \log \frac{P[x, r]}{P[x]P[r]} \tag{3.2}$$

is the mutual information between two random discrete variables X and R.[1] The idea here is that C is the maximum information that can pass through the relationship that exists between X and R. One must be free to change the alphabet and distribution of the information source X; this is a theoretical demand but also a necessity in practice. Shannon theory views information as probability distributions and capacity as a function of the conditional probability between distributions at two sites. The measure of information in a variable is the entropy of the distribution, $H(X) = -\sum_i P[x_i] \log_2 P[x_i]$ in bits. On occasion we may want to restrict the input random variable X to a certain distribution, for example the uniform one, or a certain alphabet, such as the one in binary or 4-ary modulation. Then C is called a *constrained capacity*, constrained to the imposed conditions.

When the channel input and output is continuous, Eq. (3.1) can be written as

$$C = \max_{P[x]} \left\{ \iint P[x]P[r|x] \log_2 \frac{P[r|x]}{\int P[x']P[r|x']dx'} \, dr dx \right\},$$

$$= \max_{P[x]} \left\{ \int P[x] \int P[r|x] \log_2 P[r|x] \, dr \, dx - \int P[r] \log_2 P[r] \, dr \right\}. \tag{3.3}$$

---

[1] In this chapter, upper case X etc. denotes a random variable and lower case x denotes an outcome of the variable.

We will state the AWGN capacity as a theorem and review the derivation first given by Shannon in Reference 8.

Theorem 3.1 (**AWGN Channel Capacity**) *The channel capacity of one indepen-dent AWGN channel use is*

$$C = (1/2)\log[1 + 2E_{cu}/N_o] \qquad bits/channel \ use. \qquad (3.4)$$

Shannon's main steps are as follows: The Gaussian noise distribution means that the integral over r in the first term of Eq. (3.3) reduces to

$$\int P[r|x]\log_2 P[r|x]\,dr = \int [\sqrt{1/\pi N_0}\,e^{-(r-x)^2/N_o}]\log_2[\sqrt{1/\pi N_0}\,e^{-(r-x)^2/N_o}]\,dr$$

$$= (-1/2)\log_2(2\pi e N_o/2), \qquad \text{any } x.$$

Thus, the first term is simply

$$\int P[x](-1/2)\log_2(2\pi e N_o/2)\,dx = (-1/2)\log_2(2\pi e N_o/2).$$

In the second term, $-\int P[r]\log_2 P[r]\,dr$ is the entropy of R and is greater than zero; it can be shown (see Reference [1]) that the entropy of a continuous variable with energy E_{cu} is maximum when the variable is Gaussian. This means that a Gaussian R must maximize Eq. (3.3) for any X distribution. Since X is now the difference of two Gaussians, it must also be Gaussian; that is, a Gaussian X maximizes $I(X;R)$. Since X has energy E_{cu}, it must then have distribution

$$P[x] = \sqrt{1/2\pi E_{cu}}\,e^{-x^2/2E_{cu}}.$$

Since X and η are independent, the variable R must be distributed as

$$P[r] = \sqrt{1/2\pi(E_{cu} + N_o/2)}\,e^{-r^2/2(E_{cu}+N_o/2)}.$$

The second term $-\int P[r]\log_2 P[r]\,dr$ in Eq. (3.3) is then $(1/2)\log[2\pi e(E_{cu} + N_o/2)]$. The AWGN capacity is the sum of the two terms, which is Eq. (3.4). This is the solid curve in Figure 3.1.   ∎

Shannon goes on to prove that codes exist at all rates less than C bits/channel use, whose decoding error probability tends to zero as the block length grows. These code words are made up of random letters whose values are Gaussian distributed, by the distribution just given for x. What if the letters mimic simple modulation, uniformly distributed with the standard values in Section 2.2? This is a constrained capacity, that must lie at or below the AWGN capacity. Equation (3.3) can be evaluated again, this time with sums over x, the same E_{cu}, no maximizing, and $P[x]$ set to $1/M$. The codeword letters will appear to be random M-ary modulation symbols. The

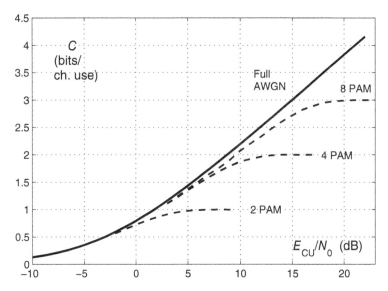

FIGURE 3.1 AWGN channel capacity in bits/channel use versus energy per channel use, compared to the capacity constrained to the 2-,4-, and 8 PAM alphabets. (4- and 8-ary data are adapted from Kressel [10], p. 111).

calculation is tedious but straightforward and the outcome is shown in Figure 3.1.[2] The capacity for an M-ary alphabet cannot exceed $\log_2 M$ bits/channel use; if the modulation alphabet is carefully chosen, C is not much less than the AWGN capacity. Otherwise there is a major loss. Above all, a small modulation alphabet must not be used at large E_{cu}/N_0 !

Extension to Time–Bandwidth. Equation (3.4) is the capacity of one independent AWGN channel. It could have come about in many ways. How many such channels exist in $\mathcal{T}$ seconds and $\mathcal{W}$ Hertz? A simple insight is provided by the pulse $\sqrt{1/T}\,\mathrm{sinc}(t/T)$ that was introduced in Chapter 1. It is T-orthogonal, has bandwidth $1/2T$ Hz, and no narrower-band orthogonal pulse exists. Consequently, a train of sinc pulses running in $\mathcal{W}$ Hz for $\mathcal{T}$ seconds provides about $\mathcal{T}/T = 2\mathcal{T}\mathcal{W}$ independent AWGN channels. In the language of Section 2.1, these each occupy a dimension of signal space.

The sinc pulse is physically unrealizable, and a precise statement of the $2\mathcal{W}\mathcal{T}$ rule needs to be hedged somewhat. A rigorous view only arrived in the 1960s with the highly technical approach of Landau, Pollak, and Slepian [11,12]. A textbook discussion appears in Reference 6, Section 5.1, and we will return to the subject in Chapter 7. For now the following simplified approach will suffice.

---

[2]The same calculation is shown in Reference 13, Figure 3.18, for in-phase and quadrature modulation via QAM. QAM is equivalent to two PAM uses.

Theorem 3.2 (**The** $2WT$ **Theorem**) *Let S be the set of all signals band limited to $(-\mathcal{W}, \mathcal{W})$ Hz and time limited to $(-\mathcal{T}/2, \mathcal{T}/2)$ in the sense that the true signals and the limited signals do not differ in energy by more than ϵ in their spectrum and time evolution. Let $\mathcal{D}$ be the dimension of S in the sense that a basis exists that represents every function $f(t)$ in S to within energy δ in $(-\mathcal{T}/2, \mathcal{T}/2)$. Then for every $\delta > \epsilon > 0$*

$$\lim_{\mathcal{T} \to \infty} \frac{\mathcal{D}}{2\mathcal{W}\mathcal{T}} \quad \to \quad 1. \tag{3.5}$$

With the theorem in hand, we can write $C_{\mathcal{W}}$, the capacity in bits/s available from $\mathcal{W}$ positive Hz, as

$$C_{\mathcal{W}} = (1/2)\log[1 + 2E_{\text{cu}}/N_o] \text{ (bits/channel use)} \times 2\mathcal{W} \text{ (uses/s)}$$
$$= \mathcal{W}\log[1 + 2E_{\text{cu}}/N_o] \qquad \text{bits/s.} \tag{3.6}$$

On a per-Hertz basis, this is

$$\log[1 + 2E_{\text{cu}}/N_o] \qquad \text{bits/Hz-s.} \tag{3.7}$$

The formula should be viewed as asymptotic, in the limit of time. To distinguish it from the single-use AWGN, it will be called the bandwidth Gaussian channel capacity.

The formula can also be expressed in terms of signal power P. In Shannon analysis, the signal is a Gaussian random process with two-sided power spectral density $S_{xx}(f)$ watts/Hz. For now we take $S_{xx}(f)$ to be constant in f. The signal power in watts, counting frequencies on $[-\mathcal{W}, \mathcal{W}]$, is therefore $P = 2\mathcal{W}S_{xx}(f)$ watts. The energy per channel use over time $\mathcal{T}$ is

$$E_{\text{cu}} = \frac{P\mathcal{T}}{2\mathcal{W}\mathcal{T}} = P/2\mathcal{W} \quad \text{J.}$$

Substitution of this into Eq. (3.6) gives the alternate capacity form

$$C_{\mathcal{W}} = \mathcal{W}\log_2(1 + P/\mathcal{W}N_o) \qquad \text{bits/s} \tag{3.8}$$

for power P uniformly distributed over $[-\mathcal{W}, \mathcal{W}]$ Hz. Equations (3.6) and (3.8) are the traditional forms for Shannon capacity over a defined bandwidth. The bandwidth $\mathcal{W}$ need not refer to a band centered at 0 Hz as here, but can refer to any piece of width $\mathcal{W}$ positive Hz.

For our purposes, there are two shortcomings in these formulas: They do not measure energy per databit carried (E_b) and they are not expressed in the more

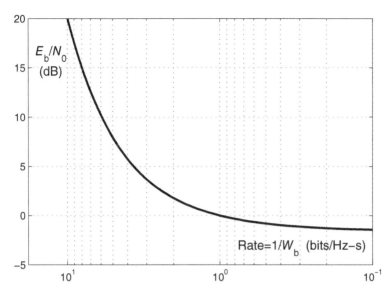

FIGURE 3.2 The bandwidth Gaussian channel Shannon limit, expressed as databit energy versus databit density in bits/Hz-s.

fundamental unit Hz-s. To achieve these aims, divide E_{cu} by the capacity in Eq. (3.4) to obtain

$$E_b/N_o = E_{cu}/CN_o = \frac{E_{cu}/N_o}{(1/2)\log(1 + 2E_{cu}/N_o)} \quad \text{J/databit,} \qquad (3.9)$$

and divide $\mathcal{W}$ by Eq. (3.6) to obtain

$$W_b = \mathcal{W}/C_{\mathcal{W}} = \frac{1}{\log(1 + 2E_{cu}/N_o)} \quad \text{Hz-s/databit.} \qquad (3.10)$$

The code rate, or *bit density* as we will call it, is $1/W_b$ databits/Hz-s. Effectively, we have here the per-databit consumption of energy and the Hz-s resource, in terms of a common parameter E_{cu}/N_0. One can combine Eqs. (3.9) and (3.10) to get

$$E_b/N_0 = (2^{1/W_b} - 1)W_b. \qquad (3.11)$$

A plot in terms of the bit density $1/W_b$ is Figure 3.2. This is a form of the *Shannon limit*: With $S_{xx}(f)$, reliable communication below and to the left of the curve is impossible.

An interesting fact is that E_b/N_0 tends to -1.59 dB as W_b grows, and reliable communication is not possible below this value. This can be shown by taking the limit in Eq. (3.11). An everyday example of the Shannon limit is given by the point (1 bit/Hz-s, 0 dB) on the curve; this states that a rate 1/2 error-correcting code combined with sinc pulse binary modulation (a 1 bit/Hz-s combination) cannot achieve arbitrarily small error probability below $E_b/N_0 = 0$ dB.

3.2   CAPACITY FOR AN ERROR RATE AND SPECTRUM

Physical transmission systems have a certain error rate and PSD, which form an important part of their specification. In what follows, we extend the traditional form of capacity to include these.

Extension to BER. Shannon and successors proved that a codeword set and receiver exist that operate with arbitrarily small error probability at all rates less than capacity, in the limit of many channel uses. In a 1959 paper [9], Shannon gave a method to add the constraint that the data bit error rate (BER) need only achieve $P_{\text{ber}} > 0$. Suppose that we have available the usual error-free channel with capacity $C^\dagger$ bits/s. Then Shannon's rate–distortion theory in Reference [9] shows that source codes exist that represent C_{ber} bits/s with accuracy P_{ber}. The two rates are related by

$$C_{\text{ber}} = C^\dagger /[1 - h_B(P_{\text{ber}})] \quad \text{bits/s.} \tag{3.12}$$

Here, C_{ber} is the nominal design rate and $C^\dagger$ is the actual rate needed in the (error-free) channel.

By following the steps that led to Eqs. (3.9) and (3.10), we obtain the per databit energy and bandwidth subject to the BER condition

$$E_b/N_o\big|_{P_{\text{ber}}} = \frac{[1 - h_B(P_{\text{ber}})]E_{\text{cu}}/N_o}{(1/2)\log(1 + 2E_{\text{cu}}/N_o)} \qquad \text{J/bit,} \tag{3.13}$$

$$W_b\big|_{P_{\text{ber}}} = \frac{[1 - h_B(P_{\text{ber}})]}{\log(1 + 2E_{\text{cu}}/N_o)} \qquad \text{Hz-s/bit.} \tag{3.14}$$

The new E_b/N_o and W_b satisfy

$$E_b/N_o = (2^{[1 - h_B(P_{\text{ber}})]/W_b} - 1)W_b. \tag{3.15}$$

At useful BER the difference in these values from those in Eqs. (3.10) and (3.11) is small. For example, at BER 0.0001, E_b/N_0 and W_b are reduced by only $1 - h_B(0.0001) = 0.9985$. At high BER, however, the differences are significant, and transmission systems can appear to violate capacity if the BER correction is not applied. The correction gives the leftward bend that appears in the next two figures.

Extension to an Arbitrary PSD. Realizable modulations do not produce signal sets with the sinc pulses and square PSD that are envisioned in Eqs. (3.6)–(3.8). This turns out to be a serious shortcoming. It might be expected that the square-PSD capacity provides a sufficiently good benchmark when the square bandwidth is equated to the set's half-power or other bandwidth measure, but this is not so as the signal databit density grows. A practical scheme's capacity reference needs to be computed for the PSD that it actually has.

The signal PSD can be incorporated as follows. To begin we need a bandwidth criterion, a concept that was introduced in Section 1.4. Simply measuring the largest bandwidth does not work well, because everyday stop band characterizations, for example, a 4-pole filter, have infinite support, at least in theory. In most of this book we use the half-power bandwidth criterion; among other reasons, all useful orthogonal pulses with symbol time T have the same half-power point, namely $1/2T$ Hz (Property 1.4).

A PSD with 1 Hz bandwidth will be taken to mean that the criterion measures 1 Hz. To work with bandwidth in a "per Hertz" sense, we will compute the capacity in bits/s for the standard 1 Hz PSD, and take the units of the result as bits/Hz-s for this PSD *shape*. Scaling this PSD width by a factor A will scale the capacity in bits/s by A but not change the capacity in bits/Hz-s.

To carry out the calculation, we return to Eq. (3.8). Let P continue as the total power but let the power density at f Hz have the shape $|H(f)|^2$, an even function with unit integral on $(-\infty, \infty)$. The PSD function at f is then $P|H(f)|^2$ watts/Hz. The power in a small piece of frequency Δf is $\approx 2P|H(f)|^2\Delta f$ watts, counting negative and positive frequencies. The quantity $W N_0$ in Eq. (3.8) becomes $\Delta f N_0$ and the capacity that results is $\Delta C = \Delta f \log_2[1 + \frac{2P}{N_0}|H(f)|^2]$. Passing to the limit $\Delta f \to df$ gives the integral

$$C_{\mathrm{psd}} = \int_0^\infty \log_2\left[1 + \frac{2P}{N_0}|H(f)|^2\right] df \quad \text{bits/Hz-s}, \tag{3.16}$$

for the total capacity of the PSD.[3] Henceforth, the subscript PSD will denote this full capacity, and the capacity C_W, divided by W, will be denoted C_{sq} as a reminder that it stems from a square PSD. If the bandwidth measure is 1 in the preceding paragraph, the units of both are bits/Hz-s. If a PSD is a factor A scaled in frequency from the standard PSD shape, the total power is scaled by A and a look at Eq. (3.16) shows that the capacity integral is scaled by A as well. Were the power not scaled by A, E_b/N_0 would not be constant. Thus the bits/Hz-s dimensions make sense.

If the PSD is square, Eqs. (3.16) and (3.8) with $W = 1$ give the same value. Occasionally, we will need C_{psd} and C_{sq} for a given PSD when the PSD's bandwidth measure is not 1. Then the dimensions of the capacities are bits/s instead of bits/Hz-s.

Several properties of C_{psd} are easily demonstrated. The first states that all useful orthogonal pulse spectra lead to higher capacity.

Property 3.1 (**Superiority of C_{psd}**) *Let the non-square PSD satisfy the spectral antisymmetry condition (Property 1.1) with T. Then*

$$C_{\mathrm{psd}} > C_{\mathrm{sq}}, \tag{3.17}$$

where C_{sq} is the capacity when the PSD is uniform on $[-1/2T, 1/2T]$ with the same total power.

---

[3] This formula appeared in Shannon's original paper [8], but applied to a different problem.

The property states that even though both spectra have the same power, the symmetric relocation of power to higher frequencies increases capacity.

Property 3.2 (**Asymptotic** C_{psd}) *Capacities for the foregoing two PSD shapes satisfy*

$$\lim_{P \to \infty} \frac{C_{\text{psd}}}{C_{\text{sq}}} = (1 + \alpha), \tag{3.18}$$

where P is total power and α is the spectrum excess bandwidth factor (as in, e.g., Eq. (1.5)).

The final property asserts that for a fixed bandwidth the square PSD is still best.

Property 3.3 *For PSDs with the same total power on a fixed bandwidth $[-W, W]$, capacity is maximized by the constant PSD.*

Calculating the Shannon Limit. Appendix 3A gives MATLAB routines for calculating the capacities and other useful quantities in this section.

Example 3.1 *(30% RC Capacities at High and Low Power)*
For a PSD that has a 30% RC shape and a half-power measure of bandwidth, we can calculate constrained capacities for total power ratio P/N_0 equal to 1 and 10,000; the last is roughly that which occurs in a local twisted-pair telephone line. The capacities are 1.001 and 16.1 bits/Hz-s (computed with gcaph in Appendix 3A). For a square PSD shape, the values are 1 and 13.3, a reduction of 2.8 bits/Hz-s at the higher power. The capacities at $P/N_0 = 10,000$ have ratio 1.21, which will grow at higher power to 1.3 according to Property 3.2. To achieve the rate 16.1 with a square PSD would require $P/N_0 = 69,800$, higher by 8.4 dB than the 30% case (found with gcapinv). This difference grows without limit as P/N_0 grows.

The standard measure of a practical system is a plot of data BER against the databit SNR, E_b/N_0. A Shannon limit in this style is obtained as follows. The PSD and the desired overall system density ρ in bits/Hz-s are fixed in advance. Repeat for each BER:

  i. *For a BER, find the actual channel rate $C^\dagger$ in bits/Hz-s. From Eq. (3.12), this is the reduced value $C^\dagger = \rho[1 - h_B(P_{\text{ber}})]$.*
  ii. *Solve the inverse capacity equation for P/N_0. For $C_{\text{psd}} = C^\dagger$, solve Eq. (3.16) for the total power ratio P/N_0 that gives $C^\dagger$.*
  iii. *Find E_b/N_0. This is simply $P/N_0 C^\dagger$.*

It is also possible to solve in the reverse direction, starting with a list of P values. This is demonstrated in Appendix 3A.

Figure 3.3 plots the outcome of procedure (*i*)–(*iii*) when the density ρ is 2 and 6 bits/Hz-s with respect to the half-power criterion and the PSD has the raised-cosine

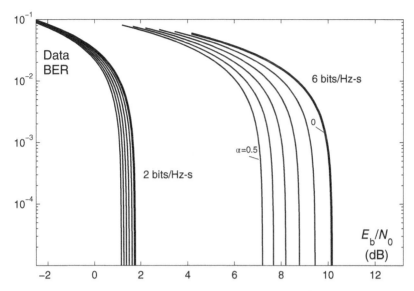

FIGURE 3.3 Shannon limit expressed as data BER versus data E_b/N_0, when the design data density is 2 and 6 bits/Hz-s. RC spectrum PSD with excess bandwidth factor $\alpha = 0, 0.1, \ldots, 0.5$.

shape with excess bandwidth $\alpha = 0, 0.1, 0.2, \ldots, 0.5$. The case $\alpha = 0$ is the square (sinc pulse) spectrum. There is little difference in the Shannon limits when the density is fixed at 2 bits/Hz-s, but 3 dB difference at 6 bits/Hz-s. Figure 3.4 fixes the RC spectrum α at 0.3, the value it will have in many examples in the book, and plots BER against E_b/N_0 at the data densities shown. The square-PSD Shannon limit is shown for the comparison. Once again, there is considerable separation between the RC and the square PSDs as the density grows.

Another bandwidth criterion such as 99% power out of band could have been used. If so, the Shannon limit in BER versus E_b/N_0 will not change, since it depends on the whole PSD, but a new databit density must be used in the calculation. How to make this change is treated in Appendix 3A and Section 6.2.

Capacity as a Limit Theorem. The Shannon theorems in this section are limit theorems: Given any spectrum shape we choose, they tell what capacity it has per second in the *limit of time*. Shannon capacity always contains such a limit. We could as well devise an alternate theory by fixing a distribution of power over time and letting bandwidth grow. We seek not to minimize time and bandwidth, but to find what flow of bits can be carried by a distribution of one in the limit of the other. In principle, one can apply the Gram–Schmidt process to a signal set occupying a certain time and bandwidth, and then find the Gaussian capacity in each dimension, but a meaningful computation would be horrendous. The "2WT" theorem leads to a way around

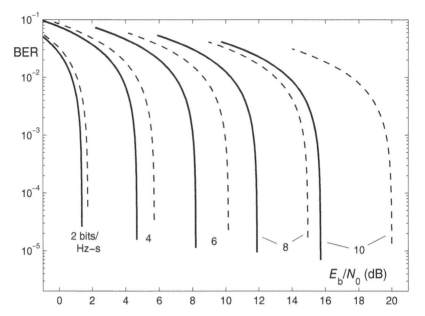

FIGURE 3.4   Shannon limit expressed as data BER versus data E_b/N_0, when $\alpha = 0.3$ with an RC spectrum. The design data density is 2–10 bits/Hz-s.

this—but it gives only an approximate answer when the time–bandwidth is finite. Still, the answer has proven good enough in a great many applications.

Another way to think about the information-carrying ability of time and bandwidth, this time not in a limit sense, is the Slepian theory in Chapter 7.

3.3   LINEAR MODULATION CAPACITY

A major issue in the Shannon theory for narrowband coding is whether restricting to simple orthogonal modulations and to the linear modulation format incurs a penalty. We would like to keep both if we can. This section will show that we should give up orthogonal transmission, but that linear modulation can reach the relevant Shannon limit when its pulse $h(t)$ is narrowband.

Linear Modulation Capacity.   Once it is modeled in discrete time, a linear modulator produces a sequence of values $b_1, b_2, \ldots$ by convolving modulator symbols $u_1, u_2, \ldots$ with a generator sequence c. How this comes to be is described in Sections 2.2–2.4. We take a communication *code* to be a set of some but not all of the possible sequences b and the codeword letters are the values b_n. The next theorem, originally due to Hirt and Massey [14,15], gives the Gaussian-channel information rate—hereafter called the capacity—of these signals in bits per use of the discrete-time channel, when the total power is P and the white Gaussian noise density is $N_0/2$. Recall that both the pulse and the modulation symbols are unit energy.

Theorem 3.3 (**ISI AWGN Capacity**) *Over an AWGN channel with noise density $N_0/2$, the maximum information rate of linear modulation with pulse $h(t)$ is*

$$C_{\text{lin}} = \int_0^{1/2T} \log_2[1 + \frac{2P}{N_0}|H_{\text{fold}}(f)|^2]\,df \qquad bits/s, \qquad (3.19)$$

where the folded spectrum is defined from the pulse spectrum $|H(f)|^2$ by Eq. (2.41) and T is the linear modulation symbol time.

Proof Sketch: Let $P(\boldsymbol{u})$ be the distribution of the modulator symbols, subject to the unit variance requirement, and take blocks of N symbols. From first principles, the constrained capacity (the maximum information rate) is

$$C_{\text{lin}} = \sup_{P(\boldsymbol{u})} \lim_{N\to\infty} (1/N)I(\boldsymbol{b}^N;\boldsymbol{u}^N)$$

$$= \sup_{P(\boldsymbol{u})} \lim_{N\to\infty} (1/N)[H(\boldsymbol{b}^N) - H(\boldsymbol{b}^N|\boldsymbol{u}^N)] \qquad bits/channel\ use.$$

Here $H(\cdot)$ is the differential entropy function. Hirt-Massey then show that the maximizing distribution $P(\boldsymbol{u})$ is Gaussian, which leads to

$$C_{\text{lin}} = T \int_0^{1/2T} \log_2\left[1 + \frac{2PT}{N_0}G(2\pi fT)\right]\,df \qquad bits/channel\ use, \quad (3.20)$$

in which $G(\cdot)$ is the Fourier transform of the autocorrelation sequence of the pulse shape $h(t)$. From the discussion at the end of Section 2.3 and Eqs. 2.42–2.44, $G(2\pi fT) = (1/T)|H_{\text{fold}}(f)|^2$. Making this substitution and dividing (3.20) by the symbol time gives (3.19). Note that the modulator alphabet is not necessarily the values of 2-ary, 4-ary, ... simple modulation; were these used, C_{lin} would be lower. Note further that when $h(t)$ is T-orthogonal, the linear modulation reverts to the bandwidth Gaussian channel envisioned in Eqs. (3.6)–(3.8).

Observe that the spectrum of the transmission is the original pulse spectrum $|H(f)|^2$, but the constrained capacity stems from the *folded* spectrum. We have immediately the following corollary:

Corollary For a pulse bandlimited to $W < 1/2T$, the capacity in Eq. (3.19) is the same form as C_{psd}, Eq. (3.16).

This follows because $|H_{\text{fold}}(f)|^2 = |H(f)|^2$ in the range of integration in Eq. (3.19). In making the calculation in Eq. (3.16), the PSD $|H(f)|^2$ is the one that actually applies to the linear modulation, not the one calibrated to unit bandwidth. The dimension thus becomes bits/s instead of bits/Hz-s.

The corollary shows that linear modulation achieves the PSD capacity when its *pulse spectrum has width less than 1/2T Hz*. The OSB model applies to precisely this case, if we are willing when necessary to make up the simple basis from pulses close to sinc pulses.

Example 3.2 *(30% Linear Modulation Capacity with Aliasing)*
Consider three linear modulations based on the same 30% root RC pulse $h(t)$; the pulse is orthogonal on the interval 1/2. The modulations have three symbol times, $T = 5/14, 5/11, 1/2$ s. These lead to first aliasing replicas centered at $1/T = 14/5, 11/5, 2$ Hz, and folding frequencies 7/5, 11/10, 1 Hz in the $|H_{\text{fold}}(f)|^2$ calculation. Figure 3.5 shows the folded spectra $|H_{\text{fold}}(f)|^2$ in the respective ranges $[0, 1/2T]$ Hz that result: For folding frequency 1.4 Hz, $|H_{\text{fold}}(f)|^2$ on $[0, 1.3]$ Hz is the undamaged RC spectrum; with folding at 1.1 Hz, $|H_{\text{fold}}(f)|^2$ on $[0, 1.1]$ Hz shows some aliasing; for folding at 1 Hz, $|H_{\text{fold}}(f)|^2$ is a square spectrum on $[0, 1]$ Hz. The square spectrum means that $h(t)$ is orthogonal for symbol time 1/2 s. The geometry is such that all three PSDs have the same total power 1 W over $[-1/2T, 1/2T]$ Hz. Applying the three folded PSDs to program gcaph in Appendix 3A with $P/N_0 = 10,000$ and the original unscaled spectra gives the capacities in Eq. (3.19): They are 16.1, 14.4, and 13.3 bits/s. C_{psd} for the original PSD, now expressed in bits/s, is the first number 16.1. As the linear modulation employs longer symbol times for the same $h(t)$, the available capacity drops, reaching finally the capacity corresponding to orthogonal transmission, 13.3 bits/s. It is not straighforward here to apply a consistent bandwidth criterion to all three spectra. If one arbitrarily takes the half-power bandwidth, the bandwidths in the figure are 1, 1.1, 1 Hz. After scaling these PSDs so that they align with 1 Hz, gcaph gives capacities 16.1, 13.1, and 13.3 bits/Hz-s. Taking the max bandwidth criterion, one gets bandwidths 1.3, 1.1, 1 Hz, and aligning these with 1 Hz leads to capacities 12.4, 13.1, and 13.3 bits/Hz-s. Observe that in both cases the product of the bandwidths and the capacities per Hz-s is the three values of C_{lin} previously found:

$$\{1, 1.1, 1\} \times \{16.1, 13.1, 13.3\} = \{1.3, 1.1, 1\} \times \{12.4, 13.1, 13.3\}$$
$$= \{16.1, 14.4, 13.3\} \qquad \text{bits/s}$$

When Linear Modulation Does Not Achieve Capacity. The linear modulation cases in the example are all based a pulse that is T-orthogonal for $T = 1/2$. The modulation symbol times $5/14, 5/11, 1/2$ can be thought of as shortenings of an orthogonal symbol time $T = 1/2$ to τT, with $\tau = 5/7, 10/11, 1$. In Section 2.5.1, this concept was called classical FTN. When the corollary to Theorem 3.3 does not hold, a number of results are known for classical FTN. The important issues are whether the capacity for a pulse nonetheless reaches C_{psd}, whether it exceeds the square PSD value C_{sq} as in Property 3.1, and how much is lost with a simple M-ary modulation alphabet. These results can be summarized as follows. Details appear in References 16 and 17.

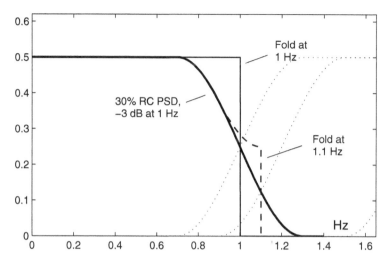

FIGURE 3.5 The 30% RC spectrum of the 1/2-orthogonal $h(t)$ (heavy line), compared to two degrees of aliasing. Dotted lines show aliasing replicas.

In all cases, $h(t)$ is a T-orthogonal pulse applied with a shorter symbol interval τT, $\tau < 1$.

- *Aliased Classical FTN.* For some τ the linear modulation capacity C_{lin} in Eq. (3.19) exceeds C_{sq} unless $h(t)$ is a sinc pulse; if h is a sinc the capacities are equal for all τ. C_{sq} in bits/s is taken here as the square-PSD capacity on bandwidth $[-1/\tau T, 1/\tau T]$ Hz with the same total power P.
- There exist $\tau < 1$ and pulses $h(t)$ for which $C_{lin} < C_{sq}$. An example is given in Reference [17]. Typically, however, C_{lin} is greater than C_{sq} and nondecreasing with decreasing τ.
- For $h(t)$ with infinite bandwidth, $C_{lin} \to C_{psd} > C_{sq}$ as $\tau \to 0$. For $h(t)$ with finite time support, there exists a power P_0 such that $C_{lin} > C_{sq}$ for all total powers $P > P_0$.
- Let $h(t)$ be bandlimited to W Hz, with $W > 1/2T$, T the symbol time; there is thus excess bandwidth. Nevertheless,

$$\frac{C_{lin}/W}{2TC_{sq}} \to 1 \quad \text{as } P \to \infty.$$

- *Simple Modulation Alphabets.* For classical FTN, a number of techniques have been developed to bound C_{lin} constrained to these alphabets [16,17]. It is known for example that the *binary* modulation C_{lin} can exceed C_{sq} with *any* alphabet for a range of practical τ and $h(t)$.

Suboptimality of Orthogonal Linear Modulation. Orthogonal modulation means that modulation symbol values are carried by T-orthogonal pulses, namely $h(t)$. By the Gibby–Smith condition, the folded spectrum of such a pulse is a constant T on $[-1/2T, 1/2T]$ Hz. Therefore, the capacity in Eq. (3.19) constrained to orthogonality is C_{sq} bits/s. When $h(t)$ is not a sinc and has excess bandwidth, the capacity constrained only to the transmitted PSD is C_{psd} bits/s, which is usually larger, and always larger when $|H(f)|^2$ satisfies the spectral antisymmetry condition (Property 1.1). As P grows, C_{psd} becomes *much* larger. In a capacity sense, orthogonal modulation can be decidedly suboptimal.

An alternate way to see the same result is to consider modulation with sinc pulses. If these carry codeword letter values, the code performance is limited to C_{sq}. Suppose the receiver and decoding are based on orthogonality. This means that only symbol-time samples of a filter matched to the orthogonal pulse are employed. Now consider an orthogonal pulse with excess bandwidth. The governing capacity is that of the new pulse, C_{psd}, but the decoder error performance cannot differ because the filter samples still have the same Gaussian statistics. *The decoding must act as if the capacity is C_{sq}.*

The fact that a significantly larger capacity can exist when $h(t)$ is not a sinc has major implications. Almost all coding schemes until now—including parity-check, convolutional, LDPC and TCM coding, whether based on binary or M-ary modulation—are based on orthogonal pulses. We need to find ways to create codes and decoders that are limited by the higher capacity. This is the aim of the next three chapters.

3.4   CONCLUSIONS

Chapter 3 has developed the Shannon capacity theory background needed for bandwidth efficient coding. The chapter began with Shannon's coding theorem and capacity for a single use of the additive white Gaussian noise channel that was introduced in Chapter 2. It broadened this result to a channel with a certain bandwidth W Hz.

The chapter expanded capacity calculation in two ways that are essential for narrowband communication: The shape of the PSD and the design data error rate were introduced as variables. These are required for a useful Shannon benchmark in the later chapters. Practical pulses have spectral sidelobes and these have a strong effect on the Shannon limit as the density of bits transmitted per Hertz and second grows.

A number of results were given for linear modulation with orthogonal and nonorthogonal pulses. For useful applications, linear modulation does not have a poorer Shannon limit in the nonorthogonal case. Orthogonal pulses often lead to significant loss and coding schemes based on them cannot achieve full capacity except with sinc pulses.

The most important results of the chapter for bandwidth efficient transmission are that orthogonal pulses should be avoided and the full PSD including side lobes must be considered.

APPENDIX 3A: CALCULATING SHANNON LIMITS

This appendix presents MATLAB functions needed for calculation of Shannon limits and capacities for the bandwidth Gaussian channel that has a given PSD, databit rate in bits/Hz-s, and databit error rate (BER). Function `rate=gcaph(p,h,f)` computes the capacity denominated in bits/Hz-s for a channel with total power p and PSD shape h that takes values on uniformly spaced non-negative frequencies f Hz, beginning at 0. The inverse function `p=gcapinv(rate,h,f)` finds the total power p that corresponds to capacity rate, for PSD h on f. Function `[eb,ber]=gbercap(rate,h,f)` solves for a list of 80 (BER, E_b/N_0) pairs suitable for plotting the Shannon limit that corresponds to rate and PSD h on f. Function `[eb,ber]=gbercapfr(rate)` extends this to the Shannon limit for the frequency FTN in Chapter 5.

Some functions are given in case they are unavailable elsewhere. These are the binary entropy function `y=hbin(x)`, its inverse function `x=hbinv(y)`, and a simple integration function `dumint(g,t)`. The last integrates $g(t)$, which takes values g on the uniformly spaced times t.

The PSD shape h should integrate to 1, or equivalently, to 1/2 on positive frequencies, but this is enforced in the program. Without loss of generality, N_0 is set to 1, and the total power P is numerically the same as P/N_0. The measure of bandwidth is determined by some feature of the PSD such as the half-power frequency—see the discussion in Chapter 1—and this feature must be aligned with 1 Hz. The rate in databits/Hz-s depends on the feature choice and is scaled accordingly to reflect 1 Hz alignment.

The choice of feature is arbitrary so long as it is consistently used, and all choices lead to the same Shannon limit. For example, a 30% root cosine spectrum can be measured by its half power frequency $1/2T$ Hz or by its highest frequency $1.3/2T$. These align 1 Hz respectively with the half-power point or the right end of the stop band. In a calculation of the Shannon limit when 2 bits are to be carried in each Hz-s, the first case requires the spectrum created by $T = 1/2$ and the rate is 2 bits/Hz-s; in the second case T must be $1.3/2$ and rate is $2/1.3$ bits/Hz-s. The rate variable reduces because the physical PSD width in the calculation is reduced by $1/1.3$; alternately, the measurement of bandwidth for each bit is inflated by 30%. Both calculations lead to the same Shannon limit plot. Examples with less simple pulses are given in Section 6.2.

Because capacity depends on the log of power, the PSD can be relatively loosely specified. However, more care with sidelobes is needed as rate grows.

Program 3A.1: Generate Points to Plot the Shannon Limit
The MATLAB function `[eb,ber]=gbercap(rate,h,f)` generates for a given design rate ρ bits/Hz-s and PSD h on f a list of BER and the corresponding E_b/N_0 in dB below which reliable communication is not possible at the BER. The function calls all the remaining functions in Appendix 3A, except Program 3A.4. The function

first computes the BER=0 power by means of the inverse capacity function gcapinv applied to ρ. The powers for the remaining BER > 0 are smaller. Rather than proceed in the inverse direction given by (i)–(iii) in Section 3.2, the program forms a list of smaller power values and works forward from these, first finding a set of error-free channel capacities $C^\dagger < \rho$ using function gcaph. These map to a BER by the inverse formula BER $= h_B^{-1}(1 - C^\dagger/\rho)$, which is performed by hbinv. The SNR E_b/N_0 is simply P/ρ. This strategy reduces the number of integrals that are evaluated.

```
function [eb,ber]=gbercap(rate,h,f)
%    Function [eb,ber]=gbercap(rate,h,f) finds the Shannon limit to BER vs Eb/No
% for signals carrying a given 'rate' bits/Hz-s, with PSD 'h' at frequencies
% 'f'=[0,...,fmax]. h is normed by the program. Calibrate h by aligning a PSD
% feature (e.g. 3 dB bandwidth) with f=1 Hz.
%  Outputs are Eb/No 'eb' in dB and achieved 'ber'
%  Requires gcaph, gcapinv, hbin, hbinv, dumint
%
pts=80;                                        %Set number of points > 20
h=h*(.5/dumint(h,f));                          %Normalize pos. PSD to 1/2
%       Find p0, highest allowed power at the nominal rate and PSD. This power
%       achieves BER=0. Uses inverse capacity function.
p0=gcapinv(rate,h,f);
  disp(['Power for BER=0 for this PSD = ',num2str(p0)])
  disp(['Eb/No(dB) for target rate = ',num2str(10*log10(p0/rate))])
%       Setup list of powers in descending tenths of a dB
p=p0*10.^(-[.0001.0003 linspace (.0007,.015,20)  .01+.01*[1:(pts-21)]]);
%       At each power in the list find the reduced 'rch' in the channel that
%       achieves BER overall. p<p0 sets rch < nominal rate; this sets BER and Eb.
rch=gcaph(p,h,f);
eb=10*log10(p./rate);                          %Eb/No for data at BER and 'rate'
ber=hbinv(1-rch/rate);                         %BER sustained by rch
```

Example 3A.1 *(Shannon limit for data rate 4 bits/Hz-s)*

For design rate $\rho = 4$ bits/Hz-s and a 30% RC spectrum, set the frequency frame to f=0:.01:1.7. The total power p for BER=0 is 11.7, numerically the same as P/N_0, which corresponds to $E_b/N_0 = P/4N_0 = 2.93$ (4.7 dB). The plot of BER against E_b/N_0 appears in Figure 3.4.

Programs 3A.2: Capacity Integral and Inverse

The function rate=gcaph(p,h,f) computes the capacity at P/N_0 ratio p and PSD shape h on f. As before, f is non-negative, uniformly spaced and starts from 0. The function sets $N_0 = 1$. If h is rescaled so that the adopted bandwidth criterion (e.g., half-power frequency) lies at 1 Hz, capacity is in units of bits/Hz-s; otherwise it is in bits/s. Integration is performed by dumint.

```
function rate=gcaph(p,h,f)
%    Function gcaph(p,h,f) computes the Gaussian capacity (bits/Hz-s) of a channel
% with total power 'p' and noise density No=1. 'f' is the pos. frequency frame
% [0,...,fmax ] and 'h' is the PSD on f with integral 1/2 and bandwidth measure 1.
% 'h' is rescaled to have power p. p and 'rate' can be can be vectors.
%  Supply dumint.
```

```
%
rate=zeros(1,length(p));
h=h/dumint(h,f);                              %Unit normalize 1-sided PSD
for i=1:length(p),                            %PSD capacity integral
   rate(i)=dumint(log2(1+p(i)*h),f);   end
```

The function p=gcapinv(r,h,f) computes the inverse function; that is, given the capacity r it finds p. It uses gcaph in a series of iterations.

```
function p=gcapinv(r,h,f)
%   Function p=gcapinv(r,h,f) performs the inverse capacity calculation. It finds
% the total power p that leads to to Gaussian capacity r bits/Hz-s with signal
% PSD h. 'f' is the frequency frame [0,...,fmax ] and 'h' is the PSD on f with
% integral 1/2 and bandwidth measure 1.
%   Supply gcaph, hbin, dumint
%
p=1; tr=gcaph(p,h,f);                         %Initial trial; p=1
%       Find rough lower and upper bounds [xd,xu] to p. 'tr' is trial
%       capacity with p.
if tr<r,                                      %p too small. Increase
   while tr<r, xd=p;
     p=2*p; tr=gcaph(p,h,f); xu=p;   end
else
   while tr>=r, xu=p;                         %p too big. Reduce.
     p=.5*p; tr=gcaph(p,h,f); xd=p;   end
end
%
%       Converge on correct p by refining lower and upper bounds.
while abs(tr-r) > .0001,                      %Test against desired rate
   p=(xd+xu)*.5; tr=gcaph(p,h,f);
   if tr>r, xu=p;                             %p too big
   else xd=p;  end                            %p too small
end
if tr>r, p=.9999*p; end                       %Be sure p is slightly small
```

Example 3A.2 *(Capacity and Its Inverse)*

With the 30% RC spectrum from Example 3.1 and total power 11.7, find capacity 4.00=gcaph(11.7,h,f). The power 11.7 corresponds to BER=0 in that example. The inverse calculation is 11.7=gcapinv(4.00,h,f). To perform the same calculations with a square PSD and the same P/N_0, set h to 1/2 on the interval [0, 1) and 0 thereafter. Then capacity is 3.675=gcaph(11.7,h,f) bits/Hz-s, a smaller rate; Eq. (3.8) yields the same with $W = 1$ and $P/N_0 = 11.7$. The inverse calculation is 11.7=gcapinv(3.675,h,f).

Programs 3A.3: Utilities

The utilities that follow are needed in the foregoing calculations. They are given in case suitable routines are not available. y=hbin(x) computes the binary entropy function $h_B(x) = -x \log_2 x - (1 - x) \log_2(1 - x)$. x=hbinv(y) is the inverse function value lying in the interval [0, 1/2]. Function int=dumint(g,t) integrates the function $g(t)$ by a simple application of Simpson's Rule. `int` is the value of the integral.

```
function y=hbin(x)
%    Function y=hbin(x) finds the binary entropy function at x. x can be
% a vector. Illegal values of x are skipped.
%
ok=find(x>0 & x<1); y=zeros(1,length(x));
if any(x<0)|any(x>1), disp(['INPUT p OUT OF RANGE IN HBIN(x)']), end
pp=x(ok); y(ok)=-pp.*log2(pp)-(1-pp).*log2(1-pp);
```

```
function x=hbinv(y)
%    Function x=hbinv(y) finds the value x in [0,.5] for which the binary
% entropy function hbin(x) is y. Precision is 10^-7. y can be a vector.
%    Supply binary entropy function 'hbin'.
%
x=zeros(1,length(y));
for k=1:length(y), yy=y(k);                             %Repeat for each y
  if yy>1 | yy<0, disp(['INPUT y OUT OF RANGE IN hbinv(y)']), return, end
  xu=.5; xd=0; tst=yy; val=hbin(tst);                   %Initialize
%          Iterate until hbin(x) is close to y.
  while abs(val-yy)>.0000001,
    if val>yy, xx=xd; xu=tst;                           %hbin(x) above yy?
    else xx=xu; xd=tst; end
    tst=tst+(yy-val)*(xx-tst)/(hbin(xx)-val+eps);       %Project next test value
    val=hbin(tst);                                      %Next trial value of x
  end
  if tst<.5, x(k)=tst; else x(k)=1-tst; end
end
```

```
function int = dumint(g,t)
%    Function int=dumint(g,t) performs a simple integral of g by summing
% up a Simpson's Rule approximation. The function is supplied by the
% values 'g' taken at evenly spaced time points 't'. The limits of
% integration are the first time point and the last.
%
n=length(g);
del=(t(n)-t(1))/(n-1);                                  %Integration subinterval
if 2*floor(n/2)==n
%          Number of values is even. Go to trapezoidal rule.
  int=del*(sum(g)-(g(1)+g(n))/2);
%          Number of values is odd. Use Simpson's Rule.
else int=(del/3)*(g(1)+g(n)+4*sum(g(2:2:n))+2*sum(g(3:2:(n-2))));
end
```

Program 3A.4: Shannon Limit for Time–Frequency FTN

For an FTN parameter combination giving `rate`, the MATLAB function `[eb,ber]=gbercapfr(rate)` estimates a list of BER and the corresponding databit E_b/N_0 in dB below which reliable communication is not possible for time–frequency FTN. This signaling is the subject of Chapter 5 and the capacity calculation method is introduced in Section 5.1. The input `rate` is $2r_{cc}/\tau\phi$, where τ is the time-FTN acceleration parameter, ϕ is the frequency FTN squeeze factor, and r_{cc} is the coding rate applied to the modulator in databits/M-ary modulator symbol. The program calls `gbercap` as a subroutine. An example is given in Section 5.1.

```
function [eb,ber]=gbercapfr(rate)
%   Function [eb,ber]=gbercapfr(rate) finds the Shannon limit to BER vs Eb/No for
% frequency FTN signals with parameter combination 2(rcc)/(tau)(phi). Set parameter
% 'rate' to this. rcc here is the coding rate in databits/M-ary modulator symbol;
% set rcc to log2 M if no coding. If no time FTN, set tau=1. Program works by calling
% gbercap for an equivalent square PSD.
%   Outputs are Eb/No 'eb' in dB and achieved 'ber'.
%   Requires gbercap, gcaph, gcapinv, hbin, hbinv, dumint

disp([' Target C = ',num2str(rate),' b/Hz-s'])
f=0:.01:1.1;                               %Frequency base
h=ones(1,length(f)); h(find(f>1))=0;       %Set h to square PSD
[eb,ber]=gbercap(rate,h,f);                %Equivalent baseband calculation
```

REFERENCES

1. *R.G. Gallager, *Information Theory and Reliable Communication*, McGraw-Hill, New York, 1968.

2. *T.M. Cover and J.A. Thomas, *Elements of Information Theory*, Wiley, New York, 1991.

3. *R.W. Yeung, *Information Theory and Network Coding*, Springer, New York, 2008.

4. *J.B. Anderson, *Digital Transmission Engineering*, 2nd ed., Wiley–IEEE Press, Piscataway, NJ, 2005.

5. *J.G. Proakis, *Digital Communication*, 4th and later eds., McGraw-Hill, New York, 1995.

6. *J.M. Wozencraft and I.M. Jacobs, *Principles of Communication Engineering*, Wiley, New York, 1965.

7. *S.G. Wilson, *Digital Modulation and Coding*, Prentice-Hall, Upper Saddle River, NJ, 1996.

8. *C.E. Shannon, Communication in the presence of noise, *Proc. IRE*, **37**, pp. 10–21, 1949; reprinted in *Claude Elwood Shannon: Collected Papers*, Sloane and Wyner, eds, IEEE Press, New York, 1993.

9. C.E. Shannon, Coding theorems for a discrete source with a fidelity criterion, *IRE Natl. Conv. Record.*, Part 4, pp. 142–163, 1959; reprinted in *Collected Papers, ibid.*

10. U. Kressel, Informationstheoretische Beurteilung digitaler Übertragungsverfahren mit Hilfe des Fehlerexponenten (Information theoretic evaluation of digital transmission methods with help of error exponents), Fortschritt-Berichte VDI, Reihe 10, Nr. 121, VDI-Verlag, Düsseldorf, 1989.

11. H.J. Landau and H.O. Pollak, Prolate spheroidal wave functions, Fourier analysis and uncertainty, III: The dimension of the space of essentially time- and bandlimited signals, *Bell Syst. Tech. J.*, **41**, pp. 1295–1336, 1962.

12. D. Slepian, On bandwidth, *Proc. IEEE*, **64**, pp. 292–300, 1976.

13. J.B. Anderson and A. Svensson, *Coded Modulation Systems*, Kluwer-Plenum, New York, 2003.

14. W. Hirt, Capacities and information rates of discrete-time channels with memory, Ph.D. thesis (no. ETH 8671), Inst. Signal and Information Processing, Swiss Fed. Inst. Tech., Zürich, 1988.

*References marked with an asterisk are recommended as supplementary reading.

15. W. Hirt and J.L. Massey, Capacity of the discrete-time Gaussian channel with intersymbol interference, *IEEE Trans. Inf. Theory*, **34**, pp. 380–388, 1988.

16. F. Rusek, Partial response and faster-than-Nyquist signaling, Ph.D. thesis, Electrical and Information Technology Dept., Lund University, Lund, Sweden, 2007.

17. F. Rusek and J.B. Anderson, Constrained capacities for faster than Nyquist signaling, *IEEE Trans. Inf. Theory*, **55**, pp. 764–775, 2009.

4

FASTER THAN NYQUIST SIGNALING

INTRODUCTION

With the fundamentals in Chapters 2 and 3 in place, it is time to look at coded faster than Nyquist signaling (FTN), the most successful coded system that works at high bits/Hz-s. The method began in the 1970s with Mazo [17], and after a slow start began to develop rapidly in the last 10–15 years. FTN is thus not at all new, but it would not have happened or even been understandable without 40 years of parallel research in Shannon capacity, equalization, coded modulation, and trellis decoding.

Coded FTN is marked by a modulator that produces a high level of intersymbol interference (ISI), an encoder such as a convolutional encoder that forms a set of code words from the modulator signals, and an iterative, or "turbo," decoder. The FTN name refers to a system of speeding up the symbol rate while keeping the same modulation pulse and spectrum, but the ideas apply to any method based on pulses with strong ISI. Chapters 2 and 3 showed that bandwidth efficiency means ISI, and that a consistent 10-dB energy saving exists between the Shannon limit and simple PAM (pulse-amplitude modulation) extremes, even at high bits/Hz-s densities. This chapter will show that practical coding schemes exist that work with the ISI and achieve most of the saving.

Section 4.1 introduces the history and notions of FTN signals. Most of the time the receiver needs to be iterative, a step up in complexity that is the penalty for near-Shannon performance. Two BCJR algorithms work together, one removing the ISI and the other decoding the code words. Since the ISI trellis description is large,

Bandwidth Efficient Coding, First Edition. John B. Anderson.
© 2017 by The Institute of Electrical and Electronics Engineers, Inc. Published 2017 by John Wiley & Sons, Inc.

some way of reducing it is necessary and Section 4.2 explores a number of these. Section 4.3 studies good convolutional codes for FTN; some ideas carry over from binary coding but some are new. Section 4.4 combines the two previous sections to form a good encoder–decoder system, and looks at decoding performance. Error rate lies several dB in energy from the true FTN Shannon limit, but occasionally touches the traditional Shannon limit for orthogonal-pulse methods. The iterative decoder's LLRs are strongly Gaussian, and by assuming that they are so, an analytical picture can be obtained in which the decoder behavior derives from the distance structure of the convolutional code and the modulator ISI.

4.1   CLASSICAL FTN

4.1.1   Origins of FTN

The faster than Nyquist approach can be said to have begun with the work of Lender and Kretzmer [5,6], which came to be called partial-response signaling (PRS). Their concept was introduction of intentional ISI in order to achieve some desired outcome, such as a more easily detected pulse or spectral zeros at critical spots, such as DC or a pilot tone location. The schemes convolved the data stream with a generator, just as in Section 2.2, but the generator taps were integers. There was no real gain in bandwidth or energy efficiency. Matters became more sophisticated in the early 1970s with the realization that ISI from whatever source could be modeled as a trellis structure [7–9].

A parallel development that began in the 1960s was the development of equalizers, that is, simple processors for removing ISI, based on filters and feedback of corrected symbols. The many variations of these are well described in communication engineering texts [1,2]. With the ISI trellis idea, it became clear that a trellis decoder could "decode" ISI just as it could its original target, convolutional codes; furthermore, it could achieve the least possible error rate, $Q(\sqrt{d_{\min}^2 E_b/N_0})$ as derived in Chapter 2. The common name for this approach was Viterbi equalizer, and a more technical name is maximum-likelihood sequence estimator (MLSE). Over the next 20 years, it became clear that the ISI trellis decoder performed far too much work, and that much reduction was possible without losing significant error rate. Apparently, the first work in this direction was Vermeulen and Hellman [12], who applied a variant of the earlier M-algorithm to the trellis. There followed many papers, but notably [11,13–16]. These investigated various aspects, and in particular Seshadri in the mid 1980s demonstrated major reductions in the decoding of severe ISI combined with coded modulations.

Classical FTN itself is a special form of PRS, and its history began in 1975 with Mazo [17]. He explored binary sinc pulse modulation, and found the surprising result that $d_{\min}$, hence the error rate, does not change when the symbol time T is reduced to τT, for τ in the range [0.802, 1]. The spectrum remains the same, a square shape on $[-1/2T, 1/2T]$ Hz, so that this is a data throughput increase of 25% above the $1/2T$ bits/s that was thought to apply. The result seemed to contradict the Nyquist limit

$1/2T$ Hz and was controversial because readers confused that limit, which applies to orthogonal pulses, with $0.802/2T$ Hz, which is the minimum bandwidth of sinc pulses with $d_{min}^2 = 2$ (details of this are in Section 2.5). The result lay mostly quiet until the 1990s, although supporting results appeared in References 18 and 19.

Nonetheless, contemporaneous studies as far back as 1979 [10] showed the variety of PRS bandwidths and minimum distances that were available with PRS and made the FTN phenomenon entirely reasonable. This work culminated with the full treatment of the problem by Said [21,22], which will be featured in Chapter 7. FTN underwent a rebirth in the late 1990s. The idea was extended to non-sinc pulses in Reference 23. A similar phenomenon applies to nonbinary transmission, to pulses that are not orthogonal for any symbol time, and roughly speaking even to nonlinear modulations[15]. A second controversy arose in the 2000s when it was discovered that the Shannon capacity of some FTN signals exceeded the traditional capacity measure then in use. This was resolved when the latter was shown to be the wrong capacity to apply; details of this are in Chapter 3.

4.1.2 Definition of FTN Signaling

In classical FTN, we start with a T-orthogonal function $h(t)$ and linear modulation $\sqrt{E_s} \sum u_n h(t - nT)$, and accelerate the appearance of new pulses by $\tau < 1$, to obtain the signal

$$s(t) = \sqrt{E_s} \sum u_n h(t - n\tau T). \tag{4.1}$$

As introduced in Section 2.5.1, this is the "accelerated pulse" definition of FTN because the pulses come earlier but are otherwise unchanged. It will be the preferred view in the rest of the book. It produces a trellis-structured set of signals with a minimum distance d_{min} that satisfies $d_{min} \leq d_0$, where d_0 is the minimum distance when $\tau = 1$. The symbol time is τT. Ordinarily, τ can drop considerably below 1 before d_{min} drops below d_0; the minimum τ giving $d_{min} = d_0$ is called the *Mazo limit*.

An alternate definition of FTN, the "stretched pulse" view, fixes the symbol interval T but scales $h(t)$ longer in time by $1/\tau$. The modulator's base pulse is then $\sqrt{\tau} h(\tau t)$. Either way, the bit density of the transmission is $\log_2 M / \tau T W$ bits/Hz-s, where W is the positive-Hz bandwidth measure of the unstretched $h(t)$ and M is the alphabet size of $\{u_n\}$.[1] From Section 1.2, all T-orthogonal pulses have half-power bandwidth $1/2T$, and taking this measure of bandwidth, we can simply write

$$\text{Bit density} = 2 \log_2 M / \tau \tag{4.2}$$

for classical FTN. Both binary and 4-ary alphabets play a role in this chapter.

Discrete-time models for the transmission in Eq. (4.1) are developed in Section 2.4: When the original $h(t)$ has bandwidth W and $W < 1/2\tau T$, there is no aliasing

---

[1]The bit density concept is introduced in Section 1.3.

and the OSB model applies, otherwise another method in Chapter 2 is applied. In this chapter, we assume no aliasing and use the OSB model. This means that $h(t)$ leads to model 2.23, namely the simple form

$$s(t) = \sqrt{E_s} \sum b_n \varphi(t - n\tau T), \qquad b_n = \sum u_{n-\ell} c_\ell, \qquad (4.3)$$

where $\{c_\ell\}$ are samples of $h(t)$, each τT and $\{\varphi(t - n\tau T)\}$ is a set of suitable basis functions orthonormal on τT. The aliasing requirement means that the OSB Condition, Property 2.1, holds; in addition, it is convenient to require Nyquist's spectral antisymmetry, Property 1.1. The matched filter front end of the receiver is still Figure 2.2, with the difference that instead of the $u_n + \eta_n$ in 2.2, the filter outputs are now $b_n + \eta_n$. As discussed in Section 2.4, these will need to be allpass filtered in order to convert the OSB model phase to maximum phase. The output of that filter is time-reversed to form a minimum-phase sequence, and there is therefore a modified $b'_n + \eta_n$ that is the input to the discrete-time detector. The noise η_n is still IID Gaussian with mean zero and variance $N_0/2$.

Figure 4.1 shows the example of binary symbols $\ldots, 0, 0, 1, -1, 1, 1, 1, 0, \ldots$ transmitted by ordinary linear modulation with a 1-orthogonal 30% excess bandwidth root RC pulse $h(t)$, and the same symbols sent with $\tau = 0.703$, that is, the pulse $\sqrt{\tau} h(\tau t)$. This τ is the Mazo limit for $h(t)$, so that $d^2_{\min} = 2$ in both cases. The 30% pulses are shown dashed. In the ordinary case, these occur each integer n seconds, and in the FTN case each $0.703n$, making a waveform that is 29.7% shorter. Yet the waveforms have a similar general appearance, reflecting the fact that their long-term PSDs are the same.

Coded FTN and Iterative Receivers. FTN can be uncoded or coded, meaning that all modulated signals can be in play, or a subset of them. Some mechanism is needed to select the subset. In this chapter, it will be a convolutional code of rate r_{cc} incoming databits per M-ary symbol out. In principle, any code can be used, although virtually all encoders used with FTN-like transmission have been convolutional. The details of code selection are in Section 4.3.

Figure 4.2 shows a discrete-time diagram of coded FTN. Across the top is the signal generation: A convolutional encoder puts out symbols, converted to M-ary if needed; next comes a long interleaver; finally, the symbols are convolved with a generator c that effects Eq. (4.3). The top structure has been standard practice for many years with ISI channels, but across the bottom is something new, an iterative decoder, or "turbo equalizer," first proposed in 1995 by Douillard et al. [24]. It consists of two BCJR algorithms in feedback configuration, with interleaver and deinterleaver, so placed that one BCJR, the CC-BCJR, sees convolutional code words and the other, the ISI-BCJR, sees a sequence with ISI. (The BCJR algorithm is introduced in Section 2.2.2.) Each BCJR informs the other about its respective progress. The genius of this structure is that each BCJR has a rather simple trellis detection to perform. After some iterations, each BCJR helping the other, the two converge to a common solution that both removes the ISI and decodes the databits. In principle one could decode the

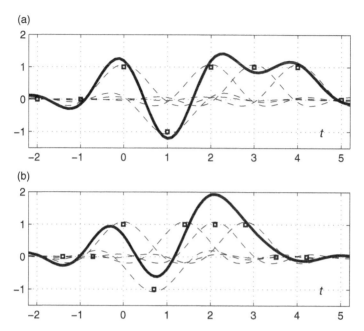

FIGURE 4.1 Comparison of ordinary linear modulation with 30% root RC pulses (a) and FTN with acceleration $\tau = 0.703$ (b). Symbol time is 1 and 0.703, respectively. Small squares show modulated values; pulses (dashed) sum to transmitted signals (heavy lines).

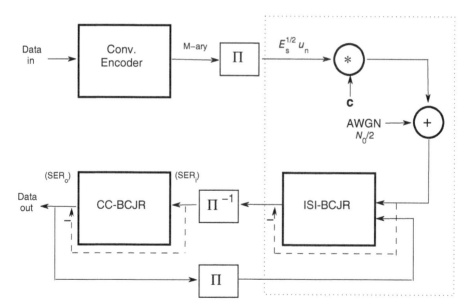

FIGURE 4.2 Discrete-time block diagram for coded FTN. Uncoded FTN includes only those parts inside the light dotted box.

signals by a traditional, hopefully ML, decoding method, but no one has yet found a way as efficient as this iterative scheme.

A critical fact about iterative decoding is that *soft* information is passed between the BCJRs, not hard symbol decisions.[2] Most often, the information is in the form of the log-likelihood ratios (LLRs) that were defined in Eqs. 2.39 and 2.40. These are essentially the probabilities of the circulating M-ary symbols. Some sort of soft processor like the BCJR is therefore required. The ISI-BCJR sees the noisy channel outputs and the soft output from the CC-BCJR; the CC-BCJR sees only the soft output of the ISI-BCJR. Another important fact is that in forming its nth output, each BCJR must de-emphasize its nth input from the other BCJR. It should focus more on inputs at other symbol positions, the so-called "extrinsic" information at position n. This is shown in Figure 4.2 as dashed line subtractions, since the job has most often been performed by subtracting the input LLR at n from the nth output LLR. This expedient is not always sufficient, but the principle holds: Proper convergence cannot occur if the BCJRs work in their strict MAP mode (Cf. Section 2.2.2).

A high-level picture of the iterative decoding is this: Convergence cannot occur until a certain minimum data E_b/N_0 is reached, called the *threshold*. Above that, the databit error rate follows that of the *convolutional code applied to a no-ISI channel with the same* E_b/N_0. But the code is working in a much narrower band channel than this, so that performance is moved much closer to the Shannon limit. This convolutional BER versus E_b/N_0 is called the *CC line*; it can be plotted by the usual methods based on code minimum distance and white noise.

Figure 4.2 is a discrete-time picture of the world as seen from the receiver. It includes the continuous-signal picture Figure 2.2, followed by any allpass filtering and a time-reversal. The choice of simple basis $\{\varphi(t - \tau T)\}$ does not appear; this appears only in Figure 2.2. M-ary modulation symbols are what circulate in the iterations. Only in the last iteration is the data decoded and presented as the receiver output. In this last iteration, both BCJRs are in strict MAP mode.

The progress of the iterations is neatly presented by a log–log plot of the M-ary symbol error probability (SER) at two points in the feedback loop, like Figure 4.3. These are the input and output error probabilities SER_i and SER_o from the CC-BCJR's point of view (for the ISI-BCJR these are simply reversed). The result is two trajectories that show all needed dynamics of the loop and are straightforward to derive from theory.[3] The first is the *CC characteristic*; it plots SER_o coming out of the CC-BCJR (the x-axis) as a function of SER_i coming into it (the y-axis). This shows the improvement in error rate realized in the convolutional decoder. The second is

---

[2] An early generation of research on iterative decoders by Jelinek and coworkers failed around 1970 primarily because feedback was not soft. This work studied concatenated codes much like the concatenated convolutions in FTN. Lack of powerful processors was also a factor. An outcome of the work was the basic BCJR paper [4].

[3] An earlier way to plot iterative decoding progress was the EXIT ("extrinsic information transfer") plot of ten Brink, which plotted symbol mutual informations instead of probabilities; for turbo equalization these are harder to compute and less informative.

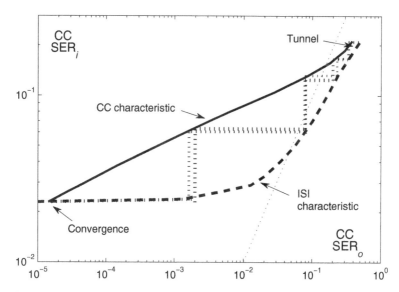

FIGURE 4.3 A typical input–output symbol error plot, showing the CC and ISI characteristics. A 100k block length leads to an iteration progress inside the heavy dotted trajectory. The CC characteristic is that of the (74,54) rate 1/2 convolutional code; the ISI characteristic is for 30% root RC, $\tau = .35$, $E_{b,\text{mod}}/N_0 = 3$ dB. (Both are discussed in succeeding sections.)

the *ISI characteristic*; it plots SER coming out of the ISI-BCJR (y-axis), that is, SER_i, as a function of SER coming in (x-axis). This is the improvement realized in the ISI decoder. The iterative process begins at the upper right in the "tunnel," and bounces from one characteristic to the other as shown in the picture until the characteristics cross, at which point the iterations stop. As interleaver block length grows, a weak law of probability applies to the characteristics and the process more and more accurately tracks a set pattern. The narrower the tunnel, the more iterations are needed to leave it, and the more are needed to reach the final cross point. If the tunnel is closed, decoding will normally fail.

There is much insight in such a plot. The CC characteristic is a straight line that depends on the code minimum distance, and a good convolutional code is one with a less negative slope, so that iterations stop at a smaller SER_o. Above all, a good code must lead to an open tunnel. The ISI characteristic can be derived from the distance structure of the ISI. It, too, should lie as far from the other characteristic as possible, with an open tunnel. There will be a new ISI characteristic for each τ and $E_{b,\text{mod}}/N_0$, where $E_{b,\text{mod}}$ is the per-bit energy of the modulator symbols. The horizontal part intersects the y-axis at the no-ISI-channel modulator performance with $E_{b,\text{mod}}/N_0$. As an aid to understanding, the plot shows a 45° light dotted line on which $\text{SER}_i = \text{SER}_o$. It can be seen that both characteristics cross this line, which means that there are some iterations during which each BCJR actually makes the SER worse; the CC-BCJR does this in early iterations and the ISI-BCJR does so in late ones. In Sections 4.2 and 4.3, we will justify all these statements.

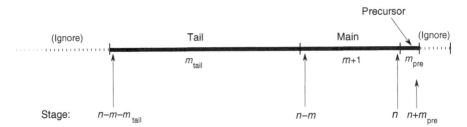

FIGURE 4.4 Discrete model regions. Lowest indexed model taps are to the right. The receiver is viewing stage n. The main model memory is m.

4.1.3 Discrete-Time Models

Because the linear modulation pulses are sufficiently narrowband, all models c in this chapter are the simple OSB type, and the base pulse is always 30% root RC. The models are listed for reference in Appendix 4A.

The modeling begins with the samples of $h(t)$ taken at $n\tau$, n an integer. As explained in Section 2.4, these are converted to maximum phase by reflecting outside the unit circle the z-plane zeros that lie inside. The reflection is performed by allpass-filtering the samples, and we are in fact free to apply any concatenation of allpasses to the samples. This is because an allpass preserves the sample autocorrelation, hence the signal spectrum and distance structure, and it keeps the noise white Gaussian with variance $N_0/2$. Finally, the sequence is reversed in time to produce the minimum-phase model. Because of the reduced ISI-BCJR, there can in fact exist a better allpass than the max-phase producing one. Once reduced, what the BCJR needs most is a steep initial rise in the tap energy, even if this must be traded for an initial section of low-energy taps. The algorithm can ignore these taps if they are less than 0.005–0.01 in absolute value and apply itself only to the main part of the pulse, which is now more tractable. The model is nearly minimum phase but not strictly so. To distinguish it from the strict case, this kind of model is called super minimum phase. More details and a sample calculation are given in Appendix 4A.

The outcome is a model consisting of a low-energy precursor with several taps, a high-energy main part, and a long decaying tail. Tail and precursor taps less than 0.005–0.01 can be ignored in the receiver and what remains is the convolutional generator model seen by the trellis detector. As a rule, very small taps can affect the signal spectrum but BCJR calculations are unaffected by them; thus they may be needed at the transmitter but they are not at the receiver.

The small-tap properties of the BCJR and the modifications that they make possible are studied in detail in Section 4.2. For the purposes there, the model can be broken down as shown in Figure 4.4 into beginning and ending regions with very small taps that can be ignored (the lengths may be zero), a precursor (whose length may also be zero), a high energy main part, and a tail. The last two will grow long as bandwidth drops. The front of the model is to the right, as it is seen by a trellis detector that traverses rightward through the trellis, and timing is shown for some trellis functions. The decision depth parameter L_D for the full-model ISI was defined in Section 2.2.2,

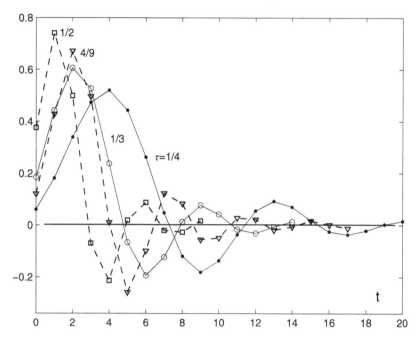

FIGURE 4.5 Some FTN discrete-time models used by the receivers in this chapter. 30% root RC base pulse. FTN accelerations are $\tau = 1/2, 4/9, 1/3, 1/4$.

and if $m + 1 + m_{\text{tail}} \geq L_D$, there should be no error performance loss at high E_b/N_0 from dropping the "ignore" taps on the left.

Figure 4.5 shows the tap response of some of the book's models, using the stretched pulse view of FTN. They have roughly the same form because they all have an RC spectrum, but they scale outward in length as τ drops. Very small precursor and tail taps are deleted in this picture so that what remains are the super minimum phase models seen at the receiver. All models have unit energy.

Figure 4.6 shows the power spectra of the full models, including the precursor taps and tail taps larger than ≈ 0.008. Frequency is normalized to the modulation symbol time and spectra have unit power. All spectra are very close in shape to RC. A log scale is used in order to show stopband side lobes, which are occasionally as large as 35 dB below the passband; they are due to truncation of the infinite response $\sqrt{\tau}h(\tau t)$ and can be lowered at the transmitter by a longer model. Some care is needed with transmitter side lobes: Lobes that build up above -20 dB can lead to misleadingly good error performance, since there is significant distance and capacity in them when bandwidth is narrow and energy is high.

4.2 REDUCED ISI-BCJR ALGORITHMS

The purpose of detection is to decide data symbols. This it will do iteratively by means of two soft-output trellis detectors that apply themselves to two trellis structures, the

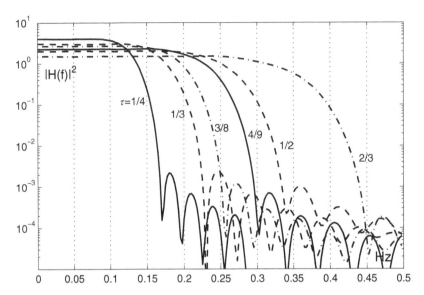

FIGURE 4.6   RC power spectra in dB versus Hz-s for discrete-time models in the chapter. τ is $1/4$–$2/3$. The RC shape is not apparent with a dB scale.

convolutional code and the ISI. The object of this section to find ways to reduce the complexity of the ISI BCJR detector. About these detectors we can make the following general comments:

- The transmission can be coded or not. If not, there is only one application of the ISI detector. It can be a BCJR, but can as well be Viterbi algorithm (VA), equalizer, or other hard-output detector.
- The object of the ISI-BCJR is to produce decisions or LLRs about M-ary modulation symbols.
- When the transmission is coded, the convolutional memory is short and the CC-BCJR is a straightforward BCJR like that in Section 2.2.2. We will not say much more about the CC-BCJR.
- The standard ISI-BCJR has forward and backward recursions, but there are simpler one-way variants and there are soft-output VAs, for example, the SOVA and Sikora/Costello algorithms [29,30].
- Almost all BCJRs in the literature work with reference to a binary-branching trellis, but both the CC-BCJR and ISI-BCJR need to be adapted to 4- and 8-branching trellises for narrowband coding. LLRs, input LLR subtraction and the backward recursion need to change.
- Some ISI models are more difficult than others. As a rule, a difficult model is one with a long length, a small $d_{\min}$, or zeros on/near the unit circle, or a combination of these. Truly narrowband transmission models are difficult to very difficult. The methods in this section are not needed with easy models.

- Computational complexity depends foremost on the number of trellis paths processed per trellis stage and their length, and much less on model length *per se*. The first object of a reduced-complexity BCJR is to reduce path numbers.

As introduced in Section 2.2.2, three major strategies for BCJR reduction are to reduce the region of calculation in the full trellis, reduce the trellis itself, or employ successive interference calculation. The nature of narrowband signals is such that all of these are effective with FTN. Only the first two are pursued here. For them, minimum phase is essential.

4.2.1   Reduced Trellis Methods: The Tail Offset BCJR

We begin with the reduced trellis approach, with reference to Figure 4.4. One can always make a hard truncation of the model. For the VA hard-decision case, the VA is applied to the trellis defined by $m + 1$ main taps only. For the BCJR soft-decision case, the standard calculation in Section 2.2.2 is applied to the same. This approach, however, ignores a certain reality: Experience shows that the size of the trellis and the accuracy of the branch labels make separate contributions to performance. Low complexity ways exist to take account of this while retaining a small trellis.

In tail offset methods, the trellis branch labels—b'_n in the BCJR Γ computation 2.32—are *offset* by a contribution from the tail taps, while the trellis size continues to be M^m. When calculation reaches trellis stage n, we write branch labels at n as

$$b'_n = \sum_{\ell=0}^{m} u_{n-\ell}\, c_\ell \; + \sum_{\ell=m+1}^{m+m_{\text{tail}}} u_{n-\ell}\, c_\ell, \tag{4.4}$$

where $\{u_n\}$ are M-ary modulation symbols and the second sum is the *label offset*. The offset is computed from tentatively decided symbols. In narrowband signaling, even a small offset can be significant, but the trellis calculation can provide reliable LLRs or hard decisions with relatively short main taps if the labels are accurate enough. The tail offset idea originated in the 1970s [25]. It was applied to VA detection of ISI by several authors in the 1980s, the best known being Duel-Hallen and Heegard [16]. They also derived a minimum distance that provides an estimate of the method's loss.

Application to VA Detection. In its work at stage n, the VA now needs to decide an earlier tentative symbol u_{n-m} that is subsequently applied to the offset generation for all trellis labels in front of it. The best procedure for the VA, used in what follows, is to associate a decision with each state at stage $n - m$; one of these is associated with each survivor after the stage n trellis extension finishes, and each leads to its own future offsets. The final symbol decision is delayed until time $n - m - m_{\text{tail}}$, and if $m + m_{\text{tail}} \geq L_D$ there is hope that the VA performance is little affected by the trellis reduction. Since the main source of VA complexity is the size of its trellis, not much is added by these associations and extra tap calculations. A tentative decision symbol is not part of the trellis state.

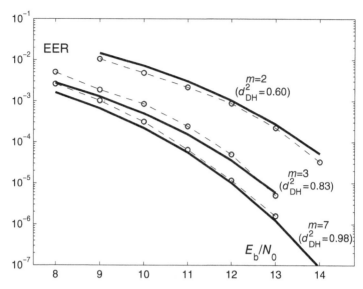

FIGURE 4.7 Tail offset VA decoding of uncoded FTN; $\tau = 1/2$ with trellis reduced to memory 2,3,7. Q-function estimate based on d_{DH} (heavy) and AWGN channel tests (circles).

Figure 4.7 shows the event error rate[4] (EER) for tail offset VA decoding of uncoded FTN transmitted over an AWGN channel. Data are taken in part from Reference [31]. The acceleration τ is 1/2 and the taps are near-minimum phase, similar to those in Appendix 4A but without the precursor. This τ causes severe ISI with 8–12 significant taps, yet major reduction in the VA computation is possible; with only 8 states (trellis memory 3) there is 0.6–0.8 dB loss in E_b/N_0 compared to a full VA.

The heavy lines in the figure show the Q-function estimate based on the Duel-Hallen–Heegard distance calculation for a state-reduced VA decoder. Their method is modified here to include the event multiplicity concept.[5] For the $m = 3, 7$ cases, the least-distant error difference event is $[2, -2, 2]$ at square distance $d_{DH}^2 = 0.83$ and 0.98, and the Q-function estimate has the form $Q(\sqrt{d_{DH}^2 E_b/N_0})/4$; the full-memory trellis has distance 1.01 and the same event. The estimate is strikingly accurate. The figure shows that much trellis reduction is possible if the labels are preserved. An insight into why is the fact that 90% of the tap energy lies in the first 4 taps, and this directs the VA search, but many more taps are needed to generate labels.

The outcome when τ is the Mazo limit 0.703 is similar, but the full minimum distance (namely 2) is reached already at $m = 2$. At $\tau = 1/4$ the distance is reached

---

[4] Techniques for EER measurement are explained in Section 4.4. The SER in the tests here is three to five times the EER. Test measurements in this chapter are based on 20–100 independent error events.

[5] Details of the DH method are in References 16 and 31. The method is similar to Program 2A.1 in Appendix 2A, but only state merges at the reduced m are considered. Multiplicity is explained in Section 2.5.1 and is employed hereafter in Q estimates.

only at about $m = 9$, which is 512 states. The VA size needs to be much larger if the model phase is not converted close to minimum phase.

Application to BCJR Detection. VA detection is hard-decision and cannot be applied to coded FTN. The BCJR produces LLR outputs and can be computed from a reduced trellis just as the VA was, with labels suitably corrected by tentative symbol decisions. A difference is that much better LLRs are produced if there is a single tail offset and it is an *expected value* computed as follows. The forward BCJR recursion at stage n produces $\alpha_n(j)$. From Eq. (2.29), this is the probability that the first n channel outputs are observed and the ISI generation is in (reduced) state j. Let $\mathcal{L}_{+1}^{n-m}$ be the set of states that have entering symbol $+1$ at stage $n - m$, and similarly for the other symbols. Then

$$\vartheta_{+1} \triangleq \sum_{j \in \mathcal{L}_{+1}^{n-m}} \alpha_n[j] \tag{4.5}$$

is the probability that the outputs are observed and $+1$ was transmitted at $n - m$. Thus the probability that the oldest main symbol is $+1$ for binary symbols is $P_{+1} = \vartheta_{+1}/(\vartheta_{+1} + \vartheta_{-1})$ and similarly for P_{-1}. Furthermore, the expected value of the symbol is

$$\mathcal{E}\{u_{n-m}\} = (+1)P_{+1} + (-1)P_{-1} = P_{+1} - P_{-1}. \tag{4.6}$$

This can be thought of as a soft symbol. The tail offset is computed as usual but from soft symbols, and there is only one set of tentative symbols for all surviving state paths. When u_{n-m} is very likely, its soft value is ± 1; when u_{n-m} is very uncertain the value is ≈ 0 and makes no contribution to the label offset.

A few details complete the description of this BCJR. In the forward recursion, no estimate of the precursor symbols is available, and the offset that might come from there is ignored. In the backward recursion, there are tentative α-based estimates, namely those in Eq. (4.5), and these can be made use of. In particular, the tail taps lie in front of the recursion and no other symbol estimates are available there. The precursor taps lie behind and can use final decisions from the BCJR λ instead. (But as a rule precursors contribute little to LLR quality). What remains is to calculate the symbol probabilities at each stage using Eq. (2.38). In uncoded FTN, the output is the most likely symbol. In iterative decoding, a somewhat modified $\sigma_k[i \rightarrow j]$ must be used in Eq. (2.38), which will be taken up at the end of this section.

Figure 4.8 compares VA to BCJR tail offset detection of uncoded binary FTN with two accelerations, $\tau = 0.35$ and $1/2$, and label offset computed from Eq. (4.6). The tap phase has a major effect on this sort of plot, and both tap sets are super minimum phase, with the phase chosen to give good SER with short memory.[6] Heavy lines show the Q-function EER estimate based on the full tap sets. As a rule, the BCJR

---

[6]The $\tau = 1/2$ set is found as in Appendix 4A. The $\tau = 0.35$, set is [0.191, 0.464, 0.623, 0.506, 0.176, −0.123, −0.196, −0.075, 0.060, 0.080, 0.013, −0.035, −0.022]; $d_{\min}^2 = 0.56$.

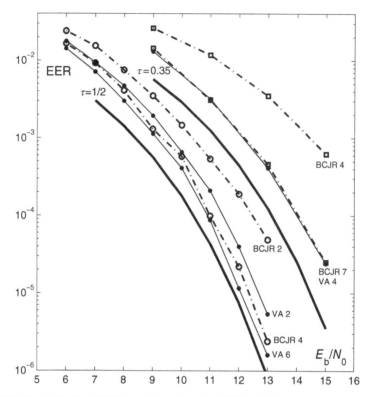

FIGURE 4.8   Tail offset BCJR decoding of uncoded FTN, with VA comparison, for $\tau = 1/2$ (squares) and 0.35 (circles). Digit shown next to VA or BCJR is the reduced state memory m. Q-function estimate (heavy line), BCJR (dash-dot), VA (thin solid line). (Data adapted from Reference 34).

needs 2–3 more memory stages than the VA to achieve the same SER when m is small. Eventually there is no further gain with m and they both approach the estimate. The least m for full BCJR performance is about 5 for $\tau = 1/2$, 6–7 for $\tau = 0.35$, and 12–13 for $\tau = 1/4$ (i.e., 4000–8000 states). These memories are the same or longer for 4-ary modulation, and the state sizes, being 4-ary, are squared. Thus the reduced-trellis approach can be attractive for binary but not for 4-ary modulation.

Other Trellis Reduction Techniques

Time Offset BCJR. We have already introduced the idea of a precursor, whose taps are at the receiver. The effect is to offset the time keeping in the trellis detector by m_{pre}, so the BCJR forward recursion first observes stage n channel outputs when the output at $n + m_{pre}$ arrives. This idea of observing at a delay can be generalized to finding the optimum offset for a given model and main observation width $m + 1$. One wants to place the main model part in the most effective trellis calculation position.

However, with a strongly minimum-phase model, the delay will be close to m_{pre}. One can also search among allpass filters to find a good combination of precursor and main part phase, which is to a degree the same problem.

Channel Shortening. Working with the first m' taps, where m' is less than the model order, without retention of an offset or other summary of earlier activity, is simple truncation of the channel model. In this scenario, it is likely that some other m' taps will work better. Channel shortening is the name that has arisen for this approach. The method began with Falconer and Magee [26] in the 1970s and has seen periodic interest since then. Better taps can be derived from minimum square-error, maximum information, minimum error rate, and other principles. Most methods place a filter or other linear processor before the receiver. The receiver is now intentionally mismatched, and so the mismatched receiver branch of coding theory leads to interesting results. A modern paper that applies prefiltering and lies relatively close to FTN signaling is Reference [27]. Direct application to coded FTN has been done by Rusek and Prlja [28,37]. A mismatched ISI model cannot ordinarily lead to ML symbol estimation, but for a given E_b/N_0 penalty one can hope for a simpler receiver, and in iterative detection one can switch to a more optimal approach late in the iterations. At this writing, it is not yet clear how the complexity of channel shortening compares to its main competitor, which is the reduced searching that comes next.

4.2.2 Reduced-Search Methods: The M-BCJR

Rather than reduce the ISI trellis size, one can search only a small part of it in the VA case or use a small part of it to calculate BCJR variables. Narrow bandwidth inevitably means high energy, and that means the contributions of almost all the trellis are vanishingly small. We begin with a review of the hard-decision uncoded PRS case, and then proceed to the soft decision BCJR, since that work is newer and iterative decoding has greater potential.

Hard Decision Reduced Searching. Reduced-search trellis decoding began with the Fano algorithm in the early 1960s and continued with various schemes to curtail the search, the best known being the M-algorithm.[7] (Note than the VA is exhaustive, not reduced.) The M-algorithm simply retains the best $\mathcal{M}$ trellis paths at each stage before continuing to the next. M-algorithm application to PRS signaling is explored in Reference [3], Section 6.6.2, and references therein; References [20] and [22] present PRS test results. For most binary PRS or ISI signals, great reduction in trellis computation is possible, often to as little as 2–3 paths, providing that the signal model is minimum phase. It is essential to so convert the signal. Analytical

---

[7]Alternate names for the M-algorithm are list decoding and beam search. In computer science, a list has an ordering, but the M-algorithm does not order paths; it retains the best $\mathcal{M}$ out of $q\mathcal{M}$ paths, where q is the branching factor of the trellis. This has order $\mathcal{M}$ computation, while maintaining a list has order $\mathcal{M} \log \mathcal{M}$. The difference is important in the sequel.

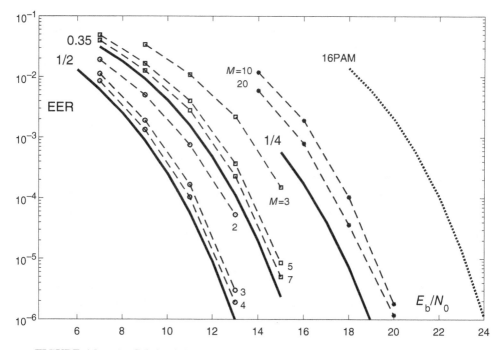

FIGURE 4.9   M-BCJR hard decoding of uncoded FTN: EER at $\mathcal{M}$ against E_b/N_0 in dB (shown dashed). From left to right, τ is $1/2$ (circles), 0.35, (squares), $1/4$ (stars). 16PAM SER is shown at the right. Q-function estimate is heavy line. (Data adapted from Reference 35.)

methods are available that relate the $\mathcal{M}$ needed for near-MLSE error performance to the distribution of signal neighbor distances. Unfortunately, the required $\mathcal{M}$ tends to grow as the PRS/ISI signal becomes more narrowband.

An M-algorithm reduced-search BCJR can be implemented by limiting the forward and backward recursions to the most promising $q\mathcal{M}$ states at each stage, where q is the trellis branching factor; only branches out of the best $\mathcal{M}$ are employed at the next stage. This will be called the *M-BCJR algorithm*. A useful view is that very small components are set to zero and the M-BCJR performs a sparse matrix calculation of Eqs. (2.33), (2.34), and (2.36). A hard-decision detector is then implemented by observing the retained set of α_n and the retained set of β_n if they overlap.[8] The calculation is easy enough with hard decisions, but challenging with soft decisions, as will be taken up presently.

Figure 4.9 shows EER versus E_b/N_0 for several $\mathcal{M}$ for uncoded FTN with $\tau = 1/2, 0.35$, and 1/4. As before, $\sqrt{\tau}h(\tau t)$ is 30% root RC and Q-function estimates are given. The maximum $\mathcal{M}$ needed for essentially MLSE error performance is only

---

[8]Specifically, if only one modulation symbol at stage $n - m$ leads to retained $\alpha_n(j)$, this is the decision. If there is no overlap of α_n and β_n at stage n, this α decision is used; if there is overlap, the usual calculation with $\sigma_k[i \to j]$ in Eq. (2.36) is made.

3, 7, and 20 paths, respectively (the closest tests to the estimates). This is considerably fewer states than the earlier reduced trellis ideas need—the $\tau = 1/4$ trellis states were more than 4000! For this case, the databit density is $2/\tau = 8$ bits/Hz-s, and the figure compares SER (the same as EER) for the simple modulation 16PAM, which has the same density. The uncoded FTN scheme has about 4-dB gain over 16PAM. VA performance is essentially the same as the largest $\mathcal{M}$ value's.

With the M-BCJR, the main model length can include the full tail, since the computational complexity is not much affected.

The ISI-BCJR Above the Mazo Limit. When the FTN τ lies above the Mazo limit, the BCJR or M-BCJR removes all the ISI in a single application in the sense that the symbol error rate is $\approx Q(\sqrt{d_0^2 E_s / N_0})$, where d_0^2 is the antipodal square distance (2 in the binary case; see Section 2.5). In an iterative receiver with intrinsic subtraction, not even perfect knowledge of other symbols can improve this. Thus only one iteration is required: One ISI-BCJR application removes the ISI, and the CC-BCJR, if any, performs decoding. The Mazo limit is therefore a boundary between two regions of receiver complexity, a single iteration being enough above the limit and full iterative decoding being needed below. In actuality, two iterations can be helpful in coded FTN, because the first leads only to error rate $AQ(\sqrt{d_0^2 E_s / N_0})$, where $A > 1$ depends on other error events with distance near d_0^2. Experiment shows that feeding the first iteration's CC-BCJR output to the ISI-BCJR reduces A to 1, which in turn improves the second CC-BCJR's data decoder output.[9]

Soft Decision M-BCJR. There are four challenging problems in the design of an M-BCJR-type algorithm that puts out accurate LLRs.

i. It easily happens that the retained $\alpha_n[j]$ at stage n do not contain any values that trace back to some of the symbols at stage $n - m$. In fact, as E_b/N_0 grows this becomes certain, since the algorithm is rightfully sure of its choice of survivors. This will force an infinite value for some of the LLRs, corresponding to a 0 or 1 probability. While a probability may be very small, iterative decoding does require values that are neither 1 nor 0. Some sort of backup value is necessary.

ii. Even if proper α values survive, it may happen that the retained α trellis states do not overlap the retained β states at some stages. Calculation of meaningful dual recursion probabilities is then impossible.

iii. A minimum-phase model is essential for efficient α computation, but to the β computation it appears as maximum phase.

iv. If the M-BCJR is part of an iterative decoder, the effect of the input *a priori* probability (AP) at stage n must be removed during the final calculation of

---

[9]The reason is that the ISI-BCJR sees essentially an antipodal decision on its second pass; see Section 4.2.3.

the stage n LLR output, the so-called intrinsic subtraction.[10] This amounts to simple subtraction of input from output LLR with binary modulation (proved at Eq. 4.9) but not with higher modulation.

All of these difficulties become more severe with narrower band signals, because energy is higher and LLRs more extreme. A simple way to solve problem (ii) is to eliminate the backward recursion, which creates a *one-way* BCJR. This generally leads to a weaker iterative decoder, but it is simpler and can be useful.

Solutions to (i). An effective way to insure that all symbols at stage $n - m$ have some probability is to distribute to them the residue α-weight that occurs when the $\mathcal{M}$ best α survivors are selected during the extension to $q\mathcal{M}$ during stage n. The forward recursion on the q extensions of the paths retained after $n - 1$ produces $q\mathcal{M}$ normalized $\alpha_n[j]$ values, of which a smaller sum remains after the selection. The residue is

$$\mathcal{R}_n = \sum_{j \in \mathcal{J}} \alpha_n[j] - \sum_{j \in \mathcal{J}_{\mathcal{M}}} \alpha_n[j], \tag{4.7}$$

where $\mathcal{J}$ and $\mathcal{J}_{\mathcal{M}}$ are the sets of indices before and after the selection. The nth residual is generally very small and is a rough estimate of probabilities of events other than the ones selected by the forward recursion; it can be distributed as needed later in the M-BCJR. The method works well with both 2- and 4-ary trellises.

Another method, described in Reference [36], carries forward a second backup recursion in case the main M-algorithm fails to include legitimate α values for all symbols. A record is kept of the tentative symbol decision path during the forward M-algorithm. In a second forward pass, all symbol extensions are forced to occur at each tentative stage and a small $\mathcal{M}$-search extends from their descendants for a few stages. Three stages and $\mathcal{M} = 2$ are usually enough. This procedure produces small, rough α-values that serve as backups. This method has so far not been shown to work with 4-ary trellises.

Solutions to (ii)–(iii). As the backward recursion progresses, the calculation (2.36) can be performed, which at stage n produces

$$\sigma_n[i \rightarrow j] = \alpha_{n-1}[i] \, \Gamma_n[i, j] \, \beta_n[j], \tag{4.8}$$

from which the LLRs may be found. The indices i and j now run over the $\mathcal{M}$ selected values. One of the above expedients will assure legal α-values for all symbols $\{u_n\}$, but it is still not guaranteed that the sets of $\mathcal{M}$ α and β states intersect, or if they do, that they will produce legal outcomes for all symbols.

There have been many attempts to relieve the intersect problem; see especially References 32–34. Among procedures that do not work well are the following:

---

[10]In the terminology of turbo decoding, intrinsic information about stage n is said to come from the AP for that stage, while extrinsic information comes from AP for the other stages.

Assigning a fixed probability ϵ when no outcome is produced for a symbol; separate forward and backward searches after which the two regions are merged into a joint structure; and employing a tail offset in the style of Duel-Hallen–Heegard instead of the expectation offset of Eq. (4.6). In early work a major shortcoming was failure to convert to minimum phase with an allpass filter; this is essential to any form of reduced search.

The first methods [35,36] that gave difficult-model results like those to come in Section 4.4 worked as follows. They performed (a) a forward α recursion with tentative symbol decisions, (b) a second forward backup recursion, and then (c) a backward β recursion. Forward recursion paths are saved for the backward recursion. The backward recursion is constrained to follow and retain all paths that it encounters whose state and stage overlap that of a stored α path, and the rest of the $\mathcal{M}$ backward paths is filled out with paths freely chosen by that recursion. The α tentative decision path is retained, and failed intersections for some symbols are repaired from backup values if necessary. Typically, there are finally only 1–3 overlaps during the backward recursion.

The description here shows some subtleties in reduced BCJR design: The β recursion lacks the direction of a minimum-phase model and must be coerced by some other means; some sort of backup is needed to supply unlikely but necessary values; the BCJR should measure good paths but it cannot simply reject bad. The scheme here will still not work well with 4-ary modulation because the natural spread in the signal probabilities is too large.

A new method proposed in Reference [38] replaces the backup recursion with the residual idea in Eq. (4.7). The residual α value is used where saved α information is insufficient. Second, the β recursion is forced to follow *all* the α paths, not just the tentative decision ones. At first this seems impossible, because the α recursion drops a fraction $(q - 1)/q$ of its paths at each stage, and the β recursion has no proper β-value to start up its recursion at the dropped trellis path ends. A way needs to be found, since all such startups do lead back to the correct trellis path. Fortunately, the BCJR recursions have the property that they rapidly "heal" an incorrect insertion. Any reasonable β-value, such as 0.01, can be placed at the dead path end, and nearly correct β-values will appear in a few stages at all $\mathcal{M}$ state nodes.[11] The $\mathcal{M}$ in the algorithm should be somewhat larger than what is needed for a forward recursion working alone. Each recursion to some degree repairs failures by the other, so that the backward recursion needs freedom to divert from what a proper smaller forward recursion would see.

Solutions to (iv). In an iterative loop with two BCJRs, experience has shown that incoming APs at stage n must not be used directly to produce the LLR outputs at the same stage. Simple subtraction of the input LLRs from the output LLRs achieves this with binary trellises for the following reason. An examination of the BCJR description in Section 2.2.2 shows that the algorithm eventually finds $P[\pm 1 \text{ sent}|r(t)]$

---

[11] The heal phenomenon is studied and measured in Reference [39]. Accurate values appear $\approx L_D$ stages later, where L_D is the decision depth of the convolutional or ISI trellis.

by evaluating Eq. (4.8) and summing over state transitions that imply ± 1. The notation "$r(t)$" refers to any and all observations that occur. The factor $\Gamma_n[i, j]$ in Eq. (4.8), defined in Eq. (2.32), takes into account the channel observation, branch label and the AP, if there is one. This AP appends an additional factor $P[u']$ in Eq. (2.32), either $P[+1]$ or $P[-1]$, and all $\sigma_n[i \rightarrow j]$ that correspond to the same symbol have the same factor. Therefore, the LLRs $\ln\left(P[+1 \text{ sent}|r(t)]/P[-1 \text{ sent}|r(t)]\right)$ in the full MAP and no-AP cases differ in that MAP appends a factor $P[+1]/P[-1]$. The incoming AP-only LLR is $\ln(P[+1]/P[-1])$, and so in summary it must be true that

$$\text{LLR}_{\text{noAP}} = \text{LLR}_{\text{MAP}} - \ln(P[+1]/P[-1]) = \text{LLR}_{\text{MAP}} - \text{LLR}_{\text{AP}}. \qquad (4.9)$$

The argument actually does not quite hold in a working receiver because incoming LLRs may be scaled there, but this "intrinsic subtraction" has proven to be a simple and effective way to prevent excessive AP propagation. Because the LLR has a different form with 4-ary symbols, subtraction does not work, and a more direct approach is needed. For a no-AP LLR, the factor $P[u']$ is simply removed from Eq. (2.32) when $\Gamma[i, j]$ is used to calculate $\sigma_n[i \rightarrow j]$. On the contrary, during the forward and backward recursions it is present.

To keep with established terminology, the proper removal of the nth AP at the final calculation of the nth LLR output will be called AP removal, intrinsic subtraction, or AP subtraction, even when there is no subtraction.

These features will construct a working M-BCJR but some theory and practice from algorithm science can make it more efficient.

- The M-algorithm has linear complexity in $\mathcal{M}$ and should be implemented as such, since $\mathcal{M}$ can be hundreds for schemes near the Shannon limit. The algorithm amounts to a search for the median of a list, a useful discussion of which appears in Knuth [40], p. 216.
- During the recursions, trellis paths in a set of $\mathcal{M}$ can have merged to the same state, and these should be combined to one path that has the sum of the respective α or β values. Merging can be ignored but the M-BCJR will act like it has a somewhat smaller $\mathcal{M}$. Linear-complexity algorithms exist to perform merging, as well as all other M-BCJR functions, and these should be used in order to preserve linear computation in the M-BCJR as a whole.
- Precision problems can destroy performance, especially with higher modulation alphabets. The heart of the matter is a computer science fact, that probabilities near 1 cannot be well enough expressed in conventional number systems (see Hayes[41]). One solution is to carry along both 1 and $1 - p$ in calculations, which is the same as carrying two LLRs instead of one (4 instead of 3 in the 4-ary case).
- As in other forms of turbo coding, the sets $\{\alpha_n\}$ and $\{\beta_n\}$ need to be normalized occasionally to unit sum, otherwise they will eventually exceed any precision.
- Rather than compute trellis branch labels during recursions, they should be precomputed in a table and looked up, using the path symbols as an address. With higher alphabets this becomes clumsy and lookups to partial tables can be combined.

4.2.3   The ISI Characteristic

Over an AWGN channel the input–output characteristic of the ISI-BCJR, reduced or not, will be similar to the dashed line in Figure 4.3. The line is determined by the ISI model and the modulator symbol energy E_s. In what follows we use ML detection and a distance analysis to show how the line arises. Two assumptions are made, that the ISI-BCJR mimics the ML detection and that the *a priori* LLRs from the CC-BCJR are approximately Gaussian. For simplicity, only binary modulation is discussed.

In coded FTN, the ISI-BCJR has as inputs the observed channel output $r(t)$ and APs from the CC-BCJR. Many researchers have observed that when represented as LLRs, these APs are roughly Gaussian in the first receiver iterations and are strongly Gaussian thereafter; the fact is easily observed by collecting statistics. Furthermore, the LLR values are independent of each other because of the interleavers. Similar conclusions hold for the CC-BCJR input. It will develop in Section 4.3.1 that these two sets of IID Gaussian AP inputs are equivalent to a second AWGN channel working on the respective code or modulator outputs. For the ISI-BCJR, it means that the BCJR sees *two* independent AWGN channels. Its behavior depends on the E_s/N_0 in the actual channel and an equivalent SNR in the AP channel.

Since the noise is assumed white Gaussian in both channels, analysis can be performed in terms of the Q-function, distance, and SNR. For the AP contribution, the BER in pictures like Figure 4.3 then translates directly to distance and SNR:

$$Q(\sqrt{2E'/N_0}) = \text{BER} = Q(\sqrt{d_{eq}^2 E_s/N_0});$$

$$\text{or} \quad d_{eq}^2 = \frac{[Q^{-1}(\text{BER})]^2}{E_s/N_0} = 2E'/E_s. \tag{4.10}$$

Here the first term in line one represents binary code word letters arriving over the AP channel with an equivalent E'/N_0 and a normalized square distance 2 between two symbol values $\pm\sqrt{E'}$. The second term equates the operand E'/N_0 to $d_{eq}^2 E_s/N_0$, where d_{eq}^2 is an *equivalent distance* with respect to the actual transmission channel and its E_s; that is, $d_{eq}^2 = 2E'/E_s$.

It remains to adapt ML symbol detection under ISI to this two-observation situation. As developed in Section 2.5, the behavior of ML detection depends on the neighbor structure of the modulator ISI trellis, and in particular it is set asymptotically in E_s/N_0 by the distance to the nearest neighbor of the transmitted trellis path. The difference now is that trellis distances are set by the composite observation of two channels. Consider a binary-alphabet error difference event Δu having L symbols, $\ell + 1$ of which are nonzero, having first symbol nonzero, and lying at square distance d_{ev}^2. The probability of deciding the entire wrong path is set by the sum of contributions from all the independent dimensions, in both channels. It is

$$Q(\sqrt{(d_{ev}^2 + \ell d_{eq}^2)E_s/N_0}). \tag{4.11}$$

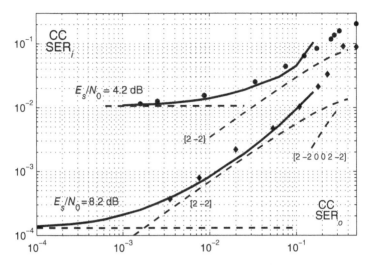

FIGURE 4.10 Construction of the ISI characteristic from difference event asymptotes, for 30% RC spectrum, FTN $\tau = 1/3$ and two E_s/N_0. Axes are standard from Figure 4.3, for which the x-axis is the *a priori* BER. The solid line is ML detection estimate composed from the significant events. Asymptotes are dashed. Points are the actual ISI characteristic with an M-BCJR and rate 2/3 convolutional code.

In constructing the AP part of Eq. (4.11), we ignore symbols where Δu is zero, since these do not contribute to the decision, and ignore its first symbol, since the iterative detector does not consider the present AP (consequently the antipodal event probability remains $Q(\sqrt{2E_s/N_0})$).

The form of Eq. (4.11) is of great interest, because it implies that the effect of AP may be characterized as a *change in distance structure*. Different *a priori* BERs can lead to different minimum-distance events. When BER ≈ 0, d_{eq}^2 is large and all difference events except the antipodal one lead to tiny probability. Hence the antipodal event has the minimum distance and the detection error probability is $Q(\sqrt{2E_s/N_0})$; this flat line is the value of the ISI characteristic as the *a priori* BER (denoted CC SER$_o$ in Figure 4.3) tends to zero. The form $Q(\sqrt{2E_s/N_0})$ states that the AP information has cleared the ISI, leaving only an antipodal decision.

A routine similar to Program 2A.3 in Appendix 2A evaluates SER and BER for a given BER in the *a priori* input. The program can use multiplicity (see Section 2.5.1) in the same way as 2A.3; the only change is the new distance in Eq. (4.11). Figure 4.10 plots the estimated ISI characteristic when the ISI stems from 30% root RC FTN with $\tau = 1/3$ and the two E_s/N_0 values 4.2 and 8.2 dB; with a rate 2/3 convolutional code these correspond to databit $E_b/N_0 = \log_{10}(3E_s/2N_0) = 6$ and 10 dB. The solid lines are the union-bound-sum estimates that significant events make to the ML-detection symbol error probability, at each *a priori* BER (the CC SER$_o$ axis).

Taking the 8.2 dB case, we see that the small-BER horizontal asymptote lies at CC $SER_i = Q(\sqrt{2E_s/N_0}) = 0.00014$. As the *a priori* BER input rises above 0.002, a new dashed asymptotic line comes into play, this one due to the difference event [2 –2]. The line is a plot of Eq. (4.11) for [2 –2] as a function of the apriori BER. At BER 0.01, the normalized square distance for [2 –2] in Eq. (4.11) is 1.54, compared to 0.72 without the help of AP information. The other error events all have higher distance than 1.54, so that event [2 –2] dominates the ML error estimate and contributes $Q(\sqrt{1.54E_s/N_0}) = .00071$ to the total 0.0008. As the *a priori* BER grows beyond 0.1 a new event [2–2 0 0 2 –2] has distance near to that of [2 –2] and begins to contribute. There are other lines nearby its line, and together they move the total ML estimate away from the [2 –2] line. The points are the true ISI characteristic with rate 2/3 binary code (iii) in Appendix 4B.[12]

The upper part of the figure shows the situation at $E_s/N_0 = 4.2$ dB, and once again error difference [2 –2] dominates at first as BER grows.

Many years' experience with reduced trellis searches like the M-algorithm show that the required $\mathcal{M}$ depends on the trellis structure's minimum distance.[13] The analysis just given shows that *a priori* BER may be characterized as a distance shift. A good BER leads to a larger minimum distance, which should lead to a small $\mathcal{M}$, and this is indeed the case. In fact, $\mathcal{M}$ can be very small in later receiver iterations, as small as 4 with the FTN τ in the range 0.4–1 and 6–8 with smaller τ.

4.3 GOOD CONVOLUTIONAL CODES

The behavior of a convolutional code during iterative decoding is specified by a straight line CC characteristic like that in Figure 4.3. The line depends to a first approximation only on the code. It can be located in the same manner as the ISI characteristic, by analyzing code behavior over an AWGN channel. The reason is the same, the input LLRs from stage to stage are nearly IID Gaussians so that the BCJR decoder acts as if it sees a certain no-ISI white Gaussian channel. This section performs the analysis and gives procedures for finding good codes for use with binary and 4-ary modulation. Those who wish only a good code can skip to the good code lists in Appendix 4B.

As in Section 4.2.3, E_s/N_0 denotes the modulator SNR, with E_s the average energy per M-ary modulation symbol. The notation E_b/N_0 is reserved for the databit SNR; with a code of rate r_{cc} databits/modulator symbol, $E_s = r_{cc}E_b$. We remind also that the CC-BCJR sees only the AP information from the ISI-BCJR, and not the channel output.

---

[12] This is rate 2/3 encoder [4 0 5; 0 4 7]. In the tests, the iteration loop is broken and CC SER_i fixed at the values in the plot by means of an equivalent AWGN channel with the same error rate. The M-BCJR in Section 4.4 is employed with $\mathcal{M}$ in the range 10–60. Block length is 150k.

[13] Such results for convolutional codes are given in Reference 44, and later in Reference 45. This behavior for the M-BCJR with FTN was first observed by Prlja [37].

In selecting a good code, there are three issues.

1. *The Usual Distance Structure.* When the last iteration is reached, at the intersection of the two characteristics in Figure 4.3, the two BCJRs switch to true MAP mode and use APs and channel observations to produce the most likely databits. Decoding is driven in the traditional way by code minimum distance and convolutional code word neighbors.

2. *The Tunnel.* No code can perform if the tunnel is closed in Figure 4.3, since decoding cannot get started. It needs to be open, and the higher the ISI characteristic, the harder this is to achieve.

3. *The Effect of AP Subtraction.* Removal of the input LLRs from the CC-BCJR output LLRs has a major effect on code behavior. Without subtraction the iterations will not converge properly, but with it decoder error is considerably worse. Other issues being equal, the code should suffer as little as possible from AP removal.

Only the first point is important in traditional decoding. The interplay of the three points depends on channel SNR, since a higher modulator E_s/N_0 pulls the ISI characteristic downward, opening the tunnel and allowing a different tradeoff of the three points. As well, points 1 and 3 combine to set the slope of the CC characteristic; the flatter the slope, the fewer the iterations and the lower the data error rate.

4.3.1   Binary CC Slope Analysis

It is possible to derive the CC characteristic slope from two code parameters, the usual free distance and a parameter ρ that measures the correlation between the code's input and output. The analysis is based on the fact that the LLRs that circulate in an iterative detector are nearly Gaussian. The slope is also easily measured from decoded data.

To start, we show that an AWGN channel is equivalent to a set of independent Gaussian LLRs. Modulate the binary convolutional code word symbols and let the modulator values be $\pm\sqrt{E_s}$. These are sent over an AWGN channel whose output is y. The decoder can observe the channel in the usual way, or it can observe the same y in the form of Λ_i, the LLR of some *a priori* information. In the second case, the LLR would be

$$\Lambda_i(y) = \ln\Big[\frac{\sqrt{1/\pi N_0}\exp(-(y-\sqrt{E_s})^2/N_0)}{\sqrt{1/\pi N_0}\exp(-(y+\sqrt{E_s})^2/N_0)}\Big] = 4y\sqrt{E_s}/N_0. \qquad (4.12)$$

Either y or Λ_i provide the same information and both are Gaussian; Λ_i has mean and variance $\mu_i = \pm 4E_s/N_0$ and $\sigma_i = \sqrt{8E_s/N_0}$. Conversely, if a set of APs are IID Gaussian, they are equivalent to an AWGN channel.

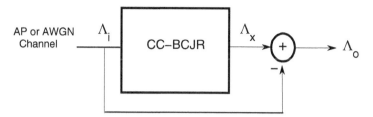

FIGURE 4.11 Definition of input, output, and AP-included LLRs.

Concerning the CC-BCJR detection, the following can be said. Whether or not its output LLR is precisely Gaussian, its SER over the AWGN channel has the asymptotic form

$$\frac{\log \text{SER}}{\log Q(\sqrt{2d_H E_s/N_0})} \to 1 \text{ as } E_s/N_0 \to \infty, \tag{4.13}$$

where d_H is the code Hamming free distance. This expression holds for ML decoding of the first symbol and asymptotically it must be the same for the CC-BCJR decision. Another fact is that as the receiver iterations come to an end, the ISI-BCJR LLR output to the CC-BCJR is Gaussian: This is because the ISI-BCJR has virtually certain knowledge of all but the present channel symbol, and the remaining symbol becomes a binary decision in Gaussian noise. The CC-BCJR is thus essentially Gaussian in its detection behavior, but it is nonetheless not certain that its output LLRs are exactly Gaussian. Still, it is widely observed during iterative decoding that input and output LLRs are nearly Gaussian. As in Section 4.2.3, we *assume* that they are Gaussian and investigate the consequences.

Define two more LLRs, Λ_o, the output with APs removed, and Λ_x, the output with APs present (see Figure 4.11). They are Gaussians with means and variances μ_o, μ_x and σ_o, σ_x. The relation $\Lambda_o = \Lambda_x - \Lambda_i$ holds, all three being Gaussian. In BCJR detection of a binary symbol, either Λ_x or Λ_o is compared to zero, and with the Gaussian assumption, the error probability is the standard form $Q(\mu/\sigma)$. This converts to a distance formulation by the substitution $\mu/\sigma = \sqrt{2d^2 E_s/N_0}$, where d functions as a equivalent operating distance that in general depends on E_s/N_0. In what follows, δ signifies the operating distance that applies with a traditional decoder that observes AP, and δ_s denotes the distance with AP absent. The subscript s signifies slope, δ_s being the inverse slope of the CC characteristic.

We also make the assumption that the covariance $\text{cov}(\Lambda_i, \Lambda_o)$ is zero; this, too, is an observed fact, for all the convolutional codes in the chapter. Finally, we define the new code parameter ρ, the correlation coefficient between Λ_x and Λ_i:

$$\rho = \frac{\mathcal{E}[\Lambda_x \Lambda_i] - \mathcal{E}[\Lambda_x]\mathcal{E}[\Lambda_i]}{\sigma_x \sigma_i}. \tag{4.14}$$

In ordinary decoding, ρ is not close to zero because the decoding performance depends on both the present symbol position (the "intrinsic information") and surrounding positions (the "extrinsic information"), and the first is significant.

Now we can prove the following:

Lemma 4.1 *Under the assumptions just given,*

$$(i)\ \text{SER} = Q(\sqrt{2\delta^2(E_s/N_0)\,E_s/N_0})\qquad AP\ allowed,$$

$$and\ \frac{\log \text{SER}}{\log Q(\sqrt{2d_H E_s/N_0})} \to 1,\quad as\ E_s/N_0 \to \infty,\qquad (4.15)$$

$$(ii)\ \text{SER} = Q(\sqrt{2\delta_s^2 E_s/N_0}),\qquad\qquad No\ AP,$$

$$where\ \delta_s^2 = [\frac{\sqrt{\delta^2(E_s/N_0)} - \rho}{\sqrt{1-\rho^2}}]^2.\qquad (4.16)$$

Furthermore, $\delta_s^2 \geq \delta^2(E_s/N_0) - 1.$ (4.17)

Proof: (i) follows from Eq. (4.13) and the assumption that Λ_x is IID Gaussian.

Moving to (ii), we note that since Λ_x is $\Lambda_o + \Lambda_i$ and Λ_o, Λ_i are assumed uncorrelated, $\mu_o = \mu_x - \mu_i$, $\sigma_o^2 = \sigma_x^2 - \sigma_i^2$, and

$$\rho = \frac{\sigma_i}{\sigma_x} + \text{cov}(\Lambda_i \Lambda_o)\frac{\sigma_o}{\sigma_x} = \frac{\sigma_i}{\sigma_x} = \frac{\sigma_i}{\sqrt{\sigma_o^2 + \sigma_i^2}}.\qquad (4.18)$$

We need an expression for σ_o and μ_o, since the ratio of these together with the Gaussian assumption gives the SER Eq. (4.16). From Eqs. (4.15) and (4.18), it must be that

$$\mu_x = \sqrt{2\delta^2 E_s/N_0}\ \sigma_x = \sqrt{2\delta^2 E_s/N_0}\ \sigma_i/\rho = \frac{4\sqrt{\delta^2}E_s/N_0}{\rho}.$$

Since $\sigma_i = \sqrt{8E_s/N_0}$, Eq. (4.18) means that

$$\sigma_o = \frac{\sqrt{1-\rho^2}}{\rho}\sqrt{8E_s/N_0}.$$

Therefore,

$$\mu_o/\sigma_o = \frac{\mu_x - \mu_i}{\sigma_o} = \frac{\sqrt{\delta^2} - \rho}{\sqrt{1-\rho^2}}\sqrt{2E_s/N_0},\qquad (4.19)$$

so that Eq. (4.16) must hold. Calculus shows that the minimum of Eq. (4.19) with respect to ρ occurs at $\rho = 1/\sqrt{\delta^2(E_s/N_0)}$, which yields Eq. (4.17). ∎

The lemma shows that there is an equivalent square-minimum distance δ^2 at each E_s/N_0, and since the lower bound (4.17) is rather tight, *removing apriori information costs about 1 unit of Hamming distance*. With the above results, the slope of the CC characteristic in Figure 4.3 is close to $1/\delta_s^2$: The log of the in-

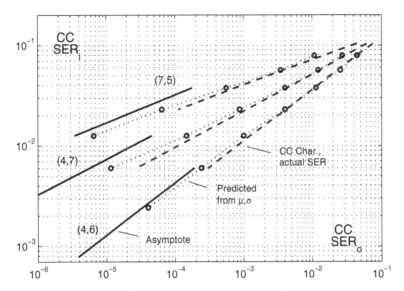

FIGURE 4.12 Plots of the CC characteristic for several rate 1/2 convolutional codes when a CC-BCJR acts alone over an AWGN channel. Shown here are actual code word symbol error rates (dashed), predicted rates based on the Gaussian LLR assumption (dotted), the same but assuming the code's d_H (solid). Code generators are defined in Appendix 4B.

coming SER_i is $\ln Q(\sqrt{2E_s/N_0}) \approx E_s/N_0$, while the log of the outgoing SER_o is $\ln Q(\sqrt{2\delta_s^2 E_s/N_0}) \approx \delta_s^2 E_s/N_0$, and so their ratio is $1/\delta_s^2$.

Figure 4.12 plots output SER versus input SER from an AWGN channel computed in three ways for some rate 1/2 convolutional codes. First, $Q(\mu/\sigma)$ is plotted from experimental input μ_i/σ_i and output μ_o/σ_o measured at a number of E_s/N_0 (the dotted lines). Second, the actual error rates for the BCJR decoder are plotted (dashed), and third, asymptotic SER_o versus SER_i from Eqs. (4.15) and (4.16) are plotted when $\delta^2(E_s/N_0)$ is set to d_H, the code Hamming distance (solid lines). The last are straight lines, but the others are not quite, since $\delta^2(E_s/N_0)$ varies somewhat with E_s/N_0. Nonetheless, all the lines in the figure are very nearly straight, a fact that simplifies searches for good codes.

The figure shows a CC-BCJR decoder working alone. When the CC-BCJR works as part of an iterative detector, its characteristic is very similar, with some variation in the upper right corner. This is a critical region, because a closed tunnel will stop the detection. To sketch the tunnel precisely enough requires a full iterative detector test, but just of the small tunnel region.

4.3.2   Good Binary Modulation Codes

This section finds good binary convolutional codes for use in FTN signaling when the code rate r_{cc} is 1/2, 2/3, and 3/4. The dimensions of r_{cc} are databits/binary modulator symbol, so the rates are numerically the same as in ordinary error-correcting coding. A higher rate generally means less complex ISI for the same transmission density in databits/Hz-s, but a more complex convolutional code.

Good codes for ordinary error correction are ones with highest Hamming free distance d_H, but codes for FTN and iterative detection have an additional parameter ρ to consider. Also, they need good tunnel performance, which is to say that they need to perform well in very bad channels; this conflicts to a degree with the best d_H and ρ. What follows is a systematic procedure to deal with these conflicting requirements, and the lists of good codes are in Appendix 4B.

The next sections assume a basic knowledge of convolutional codes at the level of Reference [1], Chapter 6. The generator notation is similar to Matlab and to the controller canonical form of Johannesson and Zigangirov [45]; that is, 2, 3, or more tapped shift register sets of length $m + 1$, whose outputs, respectively, form 2, 3, or more code word symbols (the notation is reviewed with each code rate in Appendix 4B). Minimum Hamming distances can be found using textbook methods [1,45] or standard Matlab routines. Early turbo equalization studies mostly used the rate 1/2 feed forward encoder (7,5) and sometimes the recursive systematic encoder (46,72), neither with much justification.[14] An early study of good codes was Reference [46].

First we summarize the outcome of the code searches to date. The code memory is m and the shift register(s) have $m + 1$ taps.

- To achieve a more open tunnel, m must be short and the encoder should be systematic.
- Two factors affect the CC characteristic slope, d_H and ρ. To achieve a better (i.e., smaller) slope $1/\delta_s^2$, encoders should be nonsystematic, feed forward, and have larger m.
- Once detection converges, the databit BER depends on d_H and the code column distance function. The asymptotic databit error rate has the form $Q(\sqrt{2d_H E_s/N_0}) = Q(\sqrt{2d_H r_{cc} E_b/N_0})$. From encoders with the best achievable slope at m, there normally exists one with the best d_H available at m.
- The higher code rates 2/3 and 3/4 not only reduce ISI complexity, but offer attractive structural options and in particular, more ways to achieve a systematic encoder.

There is a degree of conflict in these requirements. Much of it can be relieved by specifying the operating E_s/N_0. If it is high, the tunnel will be wide open and the best encoders are likely ones that are nonsystematic and feed forward; if it is low, a systematic encoder and a short memory will be necessary. There is also some conflict in this section with the widely held belief that convolutional encoders in turbo coding should be recursive systematic. While this may be true in ordinary turbo coding, it has not proven true in coded FTN, and recursive encoder results will not be given.

---

[14]The right-justified notation for (46,72) is (23,35), where the first entry "23" is the feedback taps; this encoder often appeared in binary turbo coding in the 1990s.

A Search Scheme. The straight line behavior of the CC characteristic makes possible an efficient search for good codes. Because long memory leads to poor tunnels, it is enough to focus on $m \leq 4$. For each type of encoder—feed forward, systematic, recursive, and so on the steps are as follows:

i. At each m, perform an exhaustive search of code generators. Form a short list of those with good slope $1/\delta_s^2$. Generators can be tested automatically by a short SER measurement with E_s/N_0 that gives SER near 0.01 and 0.08. This establishes a reliable approximate slope in a short time.

ii. Study the best-slope encoders in more detail. Using the intended ISI characteristic, check the tunnels.

iii. From the encoders with best slope, choose the generators with best d_H, best column distance, and acceptable tunnel.

Zeinali [42,43] has performed such a search of systematic and nonsystematic encoders with $m \leq 4$ at rates 1/2, 2/3, 3/4; he searched for recursive systematic encoders at rate 1/2 as well. His best encoders and some others of interest appear in Appendix 4B. We can summarize the outcome here.

Rate 1/2. At each m, the best feed forward encoders usually have better slope $1/\delta_s^2$ than the best recursive systematic ones. Extra memory often does not lead to a better CC characteristic. Systematic feed forward encoders have more open tunnels and may approach the Shannon limit more closely in its higher BER range (see Section 4.4).

Rate 2/3. Systematicity is essential for a reasonable tunnel. At rates 2/3 and 3/4, *semisystematic* encoders are available; in these some but not all of the data bits appear on the trellis branches (either 1 or 2 can appear at rate 2/3, either 1, 2, or 3 at rate 3/4). Semisystematic encoders offer an attractive combination of tunnel, slope, and d_H.

Rate 3/4. Encoders and decoders are quite complex, with memory intensive tables. The search was restricted to several hundred feed forward systematic encoders, semi and not, with good tunnel and d_H and short memory. Overall, rate 3/4 coded FTN does not perform better than rate 2/3 at the same databit density.

4.3.3 Good Convolutional Codes for 4-ary Modulation

While not what one is used to in error correction, these codes are important because tests of many code rate and FTN acceleration combinations show that 4-ary modulation is essential when many bits/Hz-s are transmitted. The code word symbols of binary convolutional codes can be collected into 4-ary modulation symbols in the obvious way, but we will show here a simpler, more direct way that at the same time has better error performance. The codes developed have rates $r_{cc} = 1$ and 4/3 databits/4-ary modulator symbol; their structures correspond to binary convolutional cousins with rates 1/2 and 2/3. As before, we need to evaluate CC characteristic slope, code minimum distance, and the tunnel size.

In order to transmit ordinary convolutional code bits by 4-ary or higher modulation, a Gray map to the 4-ary symbols is normally used. The situation is now more complicated than this simple map, because the modulator–ISI–demodulator chain and ISI-BCJR work with 4-ary symbols, denoted here $\{3, 1, -1, -3\}$, not with binary symbols. Coming out of the ISI-BCJR to a binary CC-BCJR there would need to be a 4-ary to binary LLR map; out of the CC-BCJR the reverse is needed. The mathematical conversions can be made, but there is likely to be an information loss. For example, a decision on a 4-ary symbol followed by the inverse Gray map can lead to different bits than 4-ary to binary LLR conversion followed by a decision based on the new binary LLRs.

Fortunately, there is a simple solution: The 2-bit trellis branch labels can be replaced by their corresponding Gray-mapped 4-ary symbols.[15] The result will be a convolutional code with labels taken from $\{3, 1, -1, -3\}$. Trellis decoding is the same as before except for distance, which is now Euclidean square distance, and a CC-BCJR decoder now works with 4-ary LLRs. Databits drive the encoder in the same way as before. The data BER is $\approx Q(\sqrt{d_f^2 E_b/N_0}) = Q(\sqrt{(d_f^2/r_{cc})E_s/N_0})$, where d_f is the Euclidean square free distance of the code and E_b is the energy expended per databit.

A good code of this type can be found by methods similar to those mentioned in Section 4.3.2. Now the trellis start state as well as the Gray map can affect the free distance evaluation; there are 24 possible Gray maps and 2^m start states. But longer m close the tunnel, as in Section 4.3.2, and an effective strategy is to create a short list of encoders with good d_f as in Section 4.3.2, and then evaluate CC characteristic slopes and tunnels with short tests over an AWGN channel.

An interesting property that aids in tunnel evaluation is the following. At very low databit E_b/N_0, say 1–2 dB, the convolutional decoder databit BER after 2–3 iterations is strongly predictive of how well an iterative FTN decoder will converge at this and higher E_b/N_0. This despite the fact that it is the LLR of *4-ary symbols* (whose CC-BCJR error is SER_o in the figures) that circulates in the FTN decoder. Evidently, the data BER is a better indication of the robustness of the code structure in poor channels. As before, good tunnel behavior correlates with poorer good-channel BER: A strong 4-ary code in a good channel tends to be a weak code in a bad channel. In summary, an indication of convolutional code quality in a coded FTN application can be obtained from two tests of the code alone, one in a good channel and one at $E_b/N_0 = 1$–2 dB. The two 4-ary symbol SERs show the CC characteristic slope and the weak-channel databit BER predicts the tunnel behavior.

Lists of good encoders found this way appear in Appendix 4B. An input–output correlation ρ and a lemma in the form 4.1 are problematical, because the LLR is four-dimensional and correlation depends on which of four symbols are modulated.

---

[15]Gray map hereafter means a map from q binaries to 2^q-ary modulation symbols that optimizes code Euclidean free distance. Technically, Gray map means a map with the property that a demodulator error to an adjacent 2^q-ary value causes only one databit error. This sort of map is often not optimal in coded FTN.

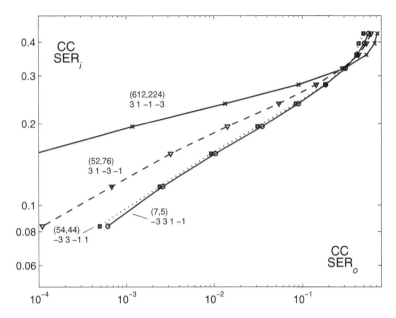

FIGURE 4.13   Plots of the CC characteristic for several rate 1 convolutional codes for use with 4-ary modulation when a CC-BCJR acts alone over an AWGN channel. Generator and Gray map notation are given in Appendix 4B.

Instead, the operating minimum distance with and without *a priori* information is given, together with the CC characteristic slope. As with binary convolutional codes, the LLRs in and out of the CC-BCJR are strongly Gaussian.

A summary of the outcome is the following:

Rate 1 Databit/4-ary Symbol Codes. Figure 4.13 shows CC characteristics of some of these encoders, measured with a CC-BCJR alone over an AWGN test channel. As appears there, the tunnel region is less favorable when the slope is better. As part of an iterative detector, the CC-BCJR has a closely matching characteristic except in the tunnel region, which is somewhat worse (lower). Input LLRs are less Gaussian there, and perhaps for this reason, the tunnel characteristic degrades. At rate 1, systematic encoders do not seem to offer improvement over nonsystematic.

Rate 4/3 Codes. The construction of these appears in Appendix 4B. By the same technique of Gray-mapping trellis branch bits directly to 4-ary symbols, rate 2/3 binary encoders can be converted to rate 4/3 databits/4-ary symbol encoders. This time the binary encoder puts out bit triples, which do not map to 4PAM directly, but two instigations of the encoder produce 6 bits, which convert to three 4-ary symbols. At high databit densities, the higher code rate is essential. For example, 6 databits/Hz-s with a rate 1 encoder requires FTN with $\tau = 1/3$, but $\tau = 4/9$ with a rate 4/3 encoder; the ISI from $\tau = 4/9$ is much easier to deal with in the ISI-BCJR. This and other rate$-\tau$ tradeoffs will be made clear in Section 4.4. The rate 4/3 codes make possible semisystematic encoders—one of each databit pair appears as a code word symbol—and these prove to be a good code type.

4.4 ITERATIVE DECODING RESULTS

In this section, we combine the previous two sections to construct an iterative decoder for coded FTN transmissions. Several major issues arise:

- How can the CC-BCJR and ISI-BCJR be best combined into a decoder? Because of intrinsic subtraction, they do not produce *a posteriori* likelihoods. On the contrary, their output LLRs are severely damaged by subtraction, and some sort of scaling or other modification is likely to aid convergence.
- As is clear in Figure 4.3, the ISI-BCJR is stronger in the early iterations, while the CC-BCJR output SER is often worse than its input SER; in the late iterations, the ISI-BCJR saturates and most of the error rate gain comes from the CC-BCJR. How should scaling reflect this?
- Different combinations of code rate r_{cc} and modulator FTN τ achieve the same transmission databits/Hz-s. Which is best?
- An FTN code/receiver combination can aim to approach the Shannon limit, which means it requires a wide open tunnel and many iterations. Or it can seek a very low data BER, which means it requires a strong code free distance, a higher E_b/N_0, and few iterations. How should the receiver design change in these two scenarios? What is the minimum computation for a given BER? How close can a combination come to Shannon?

Much research has gone into resolving these issues. Enough signal energy will make a poor design work, but what is a good design? The author, students, and colleagues have tested at least 50 receiver variations, is a process that might best be called trial and error. A consensus, however, has arisen.

A fundamental heuristic is this: *LLRs should be scaled so that they appear to come from an AWGN channel.* The motivation for this is that LLRs from stage to stage are independent (because of the interleavers) and they are evidently Gaussian. The BCJRs "think" they see an AWGN channel. But there is only one LLR scaling whose mean and variance match an AWGN channel's. This should best present the LLRs to the other BCJR.

Such an *AWGN scaling factor* can be derived as follows: Consider first binary modulation with values $\pm\sqrt{E_s}$; the received value $y = \pm\sqrt{E_s} + \eta$ itself is Gaussian with mean $\mu = \pm\sqrt{E_s}$ and variance $\sigma^2 = N_0/2$. Now suppose the LLR mean and variance are observed to be μ_o and σ_o. If these were due to an AWGN channel, LLR expression (2.40) gives the LLR value $4yE_s/N_0$, which has $\mu = \pm 4E_s/N_0$ and $\sigma^2 = (4\sqrt{E_s}/N_0)^2 N_0/2 = 8E_s/N_0$. Thus $|\sigma^2/\mu| = 2$ always, for an AWGN channel-produced LLR. Only one LLR scaling factor κ gives $|\kappa^2\sigma^2/\kappa\mu| = 2$, namely

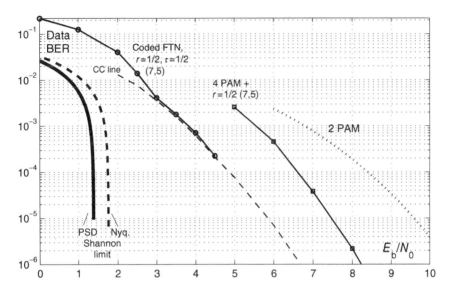

FIGURE 4.14 Shannon limits, coded binary FTN, convolutional coding + 4PAM, and uncoded 2PAM modulation, all at databit density 2 bits/Hz-s. Data BER versus data E_b/N_0. E_b is the databit energy. All systems have the same PSD.

$\kappa = 2\mu_o/\sigma_o^2$. The 4-ary modulation argument is more complex and requires numerical integration, but the outcome is approximately the same.

As the iterations progress, σ_o^2 shrinks, κ grows, and the BCJR with the larger κ dominates more. Two issues remain: LLRs are not Gaussian in the first iterations, and it is still true that one BCJR may contribute more than the other to the circulating SER. Experiment shows that both can be addressed by allowing κ to divert from its Gaussian-derived value. The best modification depends on the code and E_b/N_0, but generally speaking, κ in the first 1–3 iterations should be 1 after the ISI-BCJR (usually a higher scaling than $2\mu_o/\sigma_o^2$) and 0.4–$0.8 \times 2\mu_o/\sigma_o^2$ after the CC-BCJR (since its outputs are of poor quality).[16]

We now describe iterative receiver performance at three bit densities: 2,4, and 6 databits/Hz-s. As the density grows the outcome is markedly different: Density determines behavior. Some of the performance data have appeared in References 34–36,38,42, and 43. All ISI-BCJRs are the final M-BCJR of Section 4.2.2 with LLR scaling just discussed.

2 bits/Hz-s. Figure 4.14 compares coded FTN to other coding methods at 2 bits/Hz-s. All schemes have the same PSD. Two benchmarks delimit the figure, the Shannon limits on the left and the simple binary modulator performance on the right.

---

[16]Two further comments about scaling are (i) LLRs can scale quite large, and imposing an upper limit to the mean LLR size leads to a more stable algorithm, and (ii) scaling is more complex when E_b/N_0 nears the Shannon limit because convergence occurs in the early diagonal part of the ISI characteristic instead of the flat part.

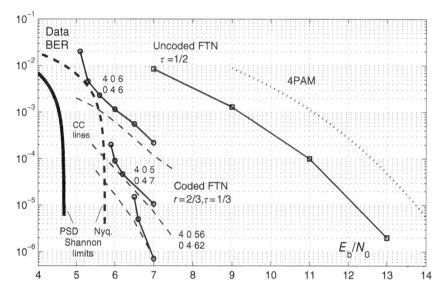

FIGURE 4.15   Shannon limits, coded and uncoded FTN, and 4PAM modulation at density 4 bits/Hz-s. Data BER versus data E_b/N_0. All systems have the same PSD.

They lie about 10 dB apart at useful BER. The two Shannon limits are for classical Nyquist and nonorthogonal signaling, as introduced in Section 3.2. FTN is not really intended for densities as low as 2, but competing methods have their best appearance there and the comparison provides valuable insight. Density 2 cannot be reached by binary modulation and ordinary error-correcting codes so that higher alphabet modulation is required with them. The least-alphabet practical example is 4PAM combined with a rate 1/2 convolutional code. The figure shows the (7,5) feed forward code, which gains 2–3 dB over the 2PAM baseline depending on the E_b/N_0; longer memory codes gain another 0.5–1 dB. Combining (7,5) with $\tau = 1/2$ FTN gains about 2 dB over over the (7,5) + 4PAM combination. Note that the coded FTN has threshold around $E_b/N_0 = 2$ dB and thereafter follows the CC line for (7,5).

4 bits/Hz-s. Figure 4.15 shows behavior at twice the previous bit density. Once again the two baselines lie about 10 dB apart at useful BER. But now the traditional Nyquist-pulse Shannon limit is 1.3 dB worse than the limit achievable with nonorthogonal pulses and the 30% RC spectrum. The two limits lie much closer at density 2; their growing separation is because more information can be carried in the PSD sidelobes. Uncoded binary FTN requires $\tau = 1/2$ and it gains about 2 dB over simple 4-ary modulation. Several coded FTN configurations are possible, for example, $r_{cc} = 1/2$, $\tau = 1/4$ or $r_{cc} = 2/3$, $\tau = 1/3$ or $r_{cc} = 3/4$, $\tau = 3/8$. Tests show that the second is the better combination, with the third a little worse. The $\tau = 1/3$ configuration touches the traditional Shannon limit and can approach the FTN limit as closely as 1–2 dB. An explanation for the $\tau = 1/3$ superiority is that the 2/3 convolutional rate is inherently stronger when both bandwidth and energy are considered and the more severe ISI with $\tau = 1/4$ is hard to equalize; as well, the

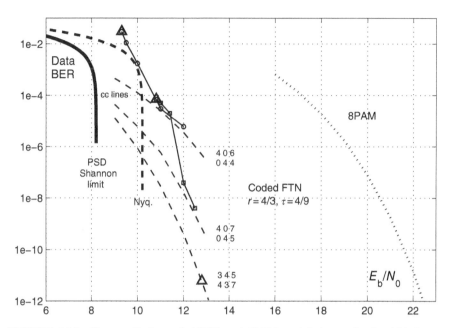

FIGURE 4.16   Shannon limits, coded FTN, and 8PAM modulation at density 6 bits/Hz-s. Data BER versus data E_b/N_0. Triangles denote approximate threshold. All systems have the same PSD.

higher rates offer more systematic and semisystematic encoder possibilities. But the easier equalization is to a degree counterbalanced by the more complex CC-BCJR. Near threshold (the leftmost point on a curve) the coded FTN systems in the figure require an M-BCJR with $\mathcal{M} = 40$–100 to start, declining to 4–8 in later iterations, and 30–80 iterations. The required block length is around 100,000. At 1–2 dB higher E_b/N_0, the $\mathcal{M}$ are much less, error rate is two decades better, and 3–6 iterations and block length 4000–8000 suffice. If there is no reason to approach capacity, the higher E_b/N_0 are much more practical.

   6 bits/Hz-s.   Figure 4.16 shows three 4-ary modulation coded FTN systems at the extreme density 6 bits/Hz-s. These employ $r_{cc} = 4/3$, $\tau = 4/9$; no schemes with the $r_{cc} = 1$, $\tau = 1/3$ combination are known that converge at reasonable E_b/N_0. The 8PAM simple modulation baseline (equivalently, 64QAM) is still about 10 dB distant but the Nyquist and FTN Shannon limits are now more than 2-dB apart. There are no competing schemes to compare to coded FTN at this density. At such high E_b/N_0 coded FTN has rather different behavior: convergence suddenly occurs at an E_b/N_0 threshold of 10–13 dB, denoted by a triangle, and the scheme then follows the CC line at a very small BER.[17] The CC lines cannot be directly tested, and instead

---

[17] The threshold and nearby performance shown is the best achieved by the 3:2 map/Gray map combinations in Appendix 4B. As with ordinary binary convolutional codes, different E_b/N_0 can lead to different best encoder designs. The CC lines themselves employ the best known maps at high E_b/N_0.

the lines are plotted from minimum distance and trellis neighbor measurements. The iteration and block length numbers are roughly those in Figure 4.15, but $\mathcal{M}$ is about twice as large.

4.5 CONCLUSIONS

Chapter 4 has introduced the faster than Nyquist signaling method. It is based on linear modulation with a strongly nonorthogonal pulse; such a pulse is essential to the narrowband signal character.

Simple modulation and the Shannon limit provide essential upper and lower benchmarks to performance of practical schemes.

While uncoded FTN can reach several dB closer to the Shannon limit than simple modulation plus coding, performance close to the limit requires coded FTN, that is, transmission with a select subset of the linear modulation signals.

Iterative decoding is essential with coded FTN, and it is necessary to reduce the complexity of the BCJR or other soft decoder that performs ISI demodulation. A number of techniques are available, the most successful of which appear to be M-BCJR algorithms and channel shortening of one kind or another.

Effective coding schemes based on convolutional coding are available, and these lead to reasonably simple decoders at 4–6 bits/Hz-s whose performance lies a few dB from the true Shannon limit. This data density is four to six times that of rate 1/2 convolutional decoding combined with binary simple modulation. Good convolutional encoders are known.

Theoretical analysis based on distance predict both the CC and ISI characteristics.

Some further comparison of FTN and earlier coding schemes will be given in Chapter 6. What can be concluded so far about coded FTN as a transmission method?

+ Coded FTN provides a method of coded transmission at 4–6 bits/Hz-s, much higher than densities commonly associated with coding.

+ The range of E_b/N_0 available between the Shannon limit and simple modulation at high densities is roughly the same as it is at more traditional densities, approximately 10 dB, and coded FTN works over most of this range.

+ The Gaussian LLR assumption provides a straighforward distance-based analysis of both the convolutional code BCJR and the ISI BCJR behavior. The analysis is not sharp enough in the tunnel region, but once the receiver leaves that region, it predicts the rest of the receiver progress.

+ Iterative decoding is essential for major E_b/N_0 gains but trellis decoding of the ISI alone is enough for gains in the 2–4 dB range.

+ Coding and complicated modulation *both* appear to be necessary for transmission that is both energy and bandwidth efficient.

APPENDIX 4A: SUPER MINIMUM-PHASE FTN MODELS

This appendix lists orthogonal simple basis (OSB) discrete-time models for most of the FTN linear modulation pulses that are used in the book. The models are obtained by sampling 30% root RC pulses $\sqrt{\tau}h(\tau t)$ for which $h(t)$ is orthogonal on the unit interval; samples smaller than ≈ 0.005 are ignored. In a process described in Sections 2.4 and 4.1, the samples are converted to a near-maximum phase sequence and then reversed to obtain a minimum-phase model.

As introduced in Sections 4.1 and 4.2, reduced-computation ISI-BCJR algorithms attain a better LLR output for a given computation size by employing a model that trades a more rapidly rising main section for a longer low-energy precursor. The BCJR ignores the precursor. Ignoring it slightly degrades the LLR quality but the main taps with energy more toward the front improve quality for a given computation limit. The trade is implemented by an allpass filter at the trellis detector input. Finding the best trade is a difficult optimization, and the models given here are only an initial suggestion. These models are said to be *super minimum phase*.

The small initial taps, if any, make up a precursor and are shown first in italics. The precursor is needed at the transmitter to insure an accurate FTN power spectrum. Figures 4.5 and 4.6 show the models and their power spectra. $d^2_{\min}$ for binary modulation is found by program `mlsedist2`; m is the model memory without the precursor; m_{ahd} is the length of the precursor and the precursor itself is shown in italics.

(i) $\tau = 1/4$ $\quad m = 23$, $m_{ahd} = 8$, $d^2_{\min} \approx .19$:
$c=[$*-.010, -.013, -.007, .005, .011, .004, -.008, .001*;
$\quad\quad$ .060, .181, .339, .473, .520, .443, .262, .047, -.120, -.182,
$\quad\quad$ -.138, -.037, .055, .092, .070, .018, -.025, -.037, -.021, .003,
$\quad\quad$ .016, .012, .0004, -.008]

(ii) $\tau = 1/3$ $\quad m = 14$, $m_{ahd} = 5$, $d^2_{\min} = .52$:
$c=[$*.016, .033, .011, -.028, .008*;
$\quad\quad$ .184, .443, .606, .527, .238, -.066, -.194, -.124, .013, .076,
$\quad\quad$ .042, -.017, -.031, -.006, .013]

(iii) $\tau = 3/8$ $\quad m = 15$, $m_{ahd} = 6$, $d^2_{\min} = .58$:
$c=[$*-.009, -.018, -.002, .013, -.003, -.002*;
$\quad\quad$ .132, .407, .624, .549, .198, -.140, -.210, -.049, .098, .087,
$\quad\quad$ -.005, -.044, -.011, .018, .008, -.009]

(iv) $\tau = 4/9$ $\quad m = 17$, $m_{ahd} = 0$, $d^2_{\min} = .86$ (2-ary), .137 (4-ary):
$c=[$*.120, .428, .671, .496, .010, -.259, -.099, .122, .083, -.056*,
$\quad\quad$ -.049, .028, .022, -.019, -.006, .017, .001, -.012]

(v) $\tau = 1/2$ $\quad m = 9$, $m_{ahd} = 8$, $d^2_{\min} = 1.01$ (2-ary), .154 (4-ary):
$c=[$*-.005, -.003, .007, -.011, -.001, .034, -.019, .003*;
$\quad\quad$ .375, .741, .499, -.070, -.214, .019, .087, -.020, -.027, .017]

(vi) $\tau = 2/3$ $\quad m = 12$, $m_{ahd} = 0$, $d^2_{\min} = 1.70$ (2-ary), .68 (4-ary):
$c=[.459, .827, .100, -.272, .124, .021, -.060, .041, -.021, .012, .003, -.009, .009]$

Example 4A.1 *(Super Min-Phase Model for $\tau = 1/2$)*

Let $h(t)$ be the 30% root RC pulse. Sampling $\sqrt{1/2}\,h(t/2)$ at $t = -20, -19, \ldots, +20$ yields 41 samples, the centermost of which are

$$\{\ldots, .040. - .109, -.053, .435, .765, .435, -.053, -.109, .040, \ldots\}. \qquad (4.20)$$

This middle phase sequence has 40 zeros total, including 19 inside the unit circle, occurring as 9 conjugate pairs, and one real zero. Reflecting these outside and time-reversing the corresponding sequence[18] produces the strict minimum-phase sequence $\{0.098, 0.408, 0.689, 0.472, -0.089, -0.279, \ldots\}$, which is plotted in Figure 4A.1. A super min-phase receiver model that leads to lower eventual error rate can be found in the following way. Find the max-phase version of a special sequence made out of just the 9 centermost samples (the values actually shown in Eq. (4.20)). The allpass that creates the new max-phase version is the second-order filter

$$B(z) = \frac{0.107 - 0.561z^{-1} + z^{-2}}{1 - 0.561z^{-1} + 0.107z^{-2}} \qquad (4.21)$$

and the time-reversed min-phase sequence is $\{0.375, 0.742, 0.500, -0.070, -0.216, \ldots\}$. This special sequence does not have an acceptable spectrum, but filtering the original sequence Eq. (4.20) with $B(z)$, or any other allpass, does produce a sequence with correct spectrum. This new time-reversed sequence has length 61; the first 18 values are very small and the next 8 are taken as the precursor; the remaining taps show a more rapid rise than the original strict min-phase sequence, as can be seen in the figure. With some small tail values deleted, the new sequence is the model in the list for $\tau = 1/2$. The point of the derivation here is to produce a new, more suitable allpass $B(z)$ for the reduced detector.

APPENDIX 4B: GOOD CONVOLUTIONAL CODES FOR FTN SIGNALING

This appendix lists good encoders at rates 1/2, 2/3, 3/4 databits/modulator symbol for binary FTN modulation and rates 1 and 4/3 for 4-ary modulation, intended for use with iterative detection in coded FTN transmission. They are found by the search method in Section 4.3.2, either by an exhaustive search for each memory and encoder type, or in the case of binary rate 3/4 and 4-ary codes, by an opportunistic random search over thousands of candidates. Code types at memory m are abbreviated by FF (feed forward), FFsys (feed-forward systematic), and FFsemi (feed-forward semisystematic). Given for each encoder are the CC characteristic slope (which drives the number of detector iterations) and the square Euclidean free distance[19] d_f^2 relative to E_b (which drives the data BER through $Q(\sqrt{d_f^2 E_b/N_0})$). This d_f^2 can be directly

---

[18]The sequence is found by Matlab function `poly`; roots are found by `roots`.

[19]The free distance of a code is the minimum distance when there is no limit to code word length. See Reference 45.

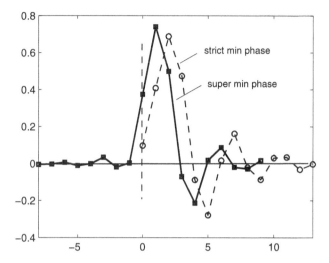

FIGURE A.1   Strict and super minimum-phase models as seen at the trellis detector. 30% root RC base pulse, $\tau = 1/2$ FTN.

compared to distances of other coding methods that lead to the same Q-function form. As always, the M-ary modulator symbol energy is $E_s = r_{cc}E_b$, and r_{cc} is the convolutional code rate in databits/M-ary symbol. In the various measurements, in this appendix, it is assumed that a CC-BCJR sees Gaussian LLRs when it works in an iterative detector and an AWGN channel when it stands alone.

For binary modulation codes, the input–output correlation parameter ρ, the CC-BCJR equivalent square distance parameter δ_s^2, and the ordinary code Hamming distance d_H are given; these are interrelated by Eqs. (4.15)–(4.17). The values of δ_s^2 and ρ are taken from SERs (rates 1/2, 2/3) or Gaussian μ/σ (rate 3/4), measured in the middle of the CC characteristic, corresponding to an SER of 0.001–0.01.

For 4-ary modulation, ρ is not available and code evaluation depends on the measured CC slope and data BER alone. The convolutional rate r_{cc} is 1 or 4/3 databits/4-ary symbol.

Binary Rate 1/2.   One databit at a time creates 2 code word bits by means of two length m tapped shift registers. Generators are written as two left-justified octals (g_1, g_2), where g_1 creates the first code word bit and g_2 the second; in the trellis description, these two bits form a branch label pair. For example, $(46, 72)$ means $(10011, 11101)$, where the left-most bits multiply mod-2 the present databit, the next-left bits multiply the previous databit, and so on, until the right-hand bits, which multiply the mth databit before the present. At rate 1/2, d_H is numerically the same as d_f^2, and the CC characteristic slope is $1/d_s^2$ given in Eq. (4.16). The data BER is

$\approx Q(\sqrt{d_f^2 E_b/N_0})$ and the code word symbol SER (no AP) is $\approx Q(\sqrt{2d_s^2 E_s/N_0})$.

(i) *FF, m* = 2:
   (7,5). Slope $1/3.9$, $\rho \approx .51$, $d_H = d_f^2 = 5$. This is the only useful $m = 2$ encoder.

(ii) *FF, m* = 3:
   (74,54), (64,54). Slope $1/5.2$, $\rho \approx .51$, $d_H = 6$.

(iii) *FF, m* = 4:
   (46,72), (66,62). Slope $1/5.2$, $\rho \approx .50$, $d_H = 7$. The larger m and d_H here do not give better slope than (ii)

(v) *FF, m* = 5:
   (62,57). Slope $1/6.5$, $d_H = 8$. Others are as good.

Binary Rate 2/3. Two databits at a time create 3 code word bits by means of six length m shift registers. Generators are written as six left-justified octals in the form $\begin{bmatrix} g_{11} & g_{12} & g_{13} \\ g_{21} & g_{22} & g_{23} \end{bmatrix}$. Here g_{ij} are the shift register taps that create the contribution to code word branch bit j from databit i. From each trellis node stem $2^2 = 4$ branches, and the 3 bits in each trellis label are the mod-2 sum of contributions from databits 1 and 2. At rate 2/3, BER and SER are $\approx Q(\sqrt{d_f^2 E_b / N_0})$ and $\approx Q(\sqrt{2 d_s^2 E_s / N_0})$ as before, but now $d_f^2 = 4 d_H / 3$.

(i) *FFsys, m* = 1:
$\begin{bmatrix} 4 & 0 & 6 \\ 0 & 4 & 6 \end{bmatrix}$ Slope $1/1.1$, $\rho \approx .68$, $d_H = 2$, $d_f^2 = 4 d_H / 3 = 8/3$. Best tunnel and best d_H for code type and memory.

(ii) *FF, m* = 2:
$\begin{bmatrix} 0 & 1 & 7 \\ 5 & 6 & 1 \end{bmatrix}, \begin{bmatrix} 5 & 1 & 6 \\ 1 & 2 & 3 \end{bmatrix}$. Slope $1/3.5$, $\rho \approx .60$, $d_H = 4$, $d_f^2 = 16/3$. Many encoders have these parameters.

$\begin{bmatrix} 3 & 4 & 5 \\ 4 & 3 & 7 \end{bmatrix}$. Slope $1/4.2$, $\rho \approx .61$, $d_H = 5$, $d_f^2 = 20/3$. Best slope and d_H for code type and m.

(iii) *FFsys, m* = 2:
$\begin{bmatrix} 4 & 0 & 5 \\ 0 & 4 & 7 \end{bmatrix}$. Slope $1/2.0$, $\rho \approx .65$, $d_H = 3$, $d_f^2 = 4$. Best slope encoder for type and m.

(iv) *FFsemisys, m* = 2:
$\begin{bmatrix} 0 & 1 & 7 \\ 4 & 6 & 1 \end{bmatrix}$. Slope $1/3.2$, $\rho \approx .62$, $d_H = 4$, $d_f^2 = 16/3$. Good d_H for type and m.

(v) *FFsys, m* = 3:

$$\begin{bmatrix} 40 & 0 & 64 \\ 0 & 40 & 74 \end{bmatrix}, \begin{bmatrix} 40 & 0 & 54 \\ 0 & 40 & 64 \end{bmatrix}.$$ Slope $1/3.8$, $\rho \approx .63$, $d_H = 4$, $d_f^2 = 16/3$. Two of best slope encoders for code type and *m*.

(vi) *FFsys, m* = 4:

$$\begin{bmatrix} 40 & 0 & 62 \\ 0 & 40 & 72 \end{bmatrix}, \begin{bmatrix} 40 & 0 & 56 \\ 0 & 40 & 62 \end{bmatrix}.$$ Slope $1/5.2$, $\rho \approx .62$, $d_H = 4$, $d_f^2 = 16/3$. Two of best slope encoders for type and *m*.

Binary Rate 3/4. Three databits at a time create 4 code word bits by means of 12 length-*m* shift registers. Generators are written as 12 left-justified octals in the form

$$\begin{bmatrix} g_{11} & g_{12} & g_{13} & g_{14} \\ g_{21} & g_{22} & g_{23} & g_{24} \\ g_{31} & g_{32} & g_{33} & g_{34} \end{bmatrix}.$$

Here, g_{ij} are defined as before, but with databit $i = 1, 2, 3$. From each trellis node stem $2^3 = 8$ branches, and the 4 bits in each trellis label are the mod-2 sum of contributions from databits 1, 2, and 3. At rate 3/4, BER and SER are the same Q forms as before, but $d_f^2 = 3d_H/2$.

(i) *FFsys, m* = 1:

$$\begin{bmatrix} 4 & 0 & 0 & 6 \\ 0 & 4 & 0 & 4 \\ 0 & 0 & 4 & 6 \end{bmatrix}.$$ Slope $1/1.0$, $\rho \approx .74$, $d_H = 2$, $d_f^2 = 3d_H/2 = 3$. Good tunnel and best d_H for code type and *m*.

(ii) *FFsemisys, m* = 1:

$$\begin{bmatrix} 4 & 0 & 2 & 6 \\ 0 & 4 & 0 & 6 \\ 0 & 0 & 6 & 4 \end{bmatrix}.$$ Slope $1/2.1$, $\rho \approx .70$, $d_H = 3$, $d_f^2 = 3$. Good tunnel and best d_H for type and *m*.

(iii) *FF, m* = 2:

$$\begin{bmatrix} 6 & 2 & 2 & 6 \\ 1 & 6 & 0 & 7 \\ 0 & 2 & 5 & 5 \end{bmatrix}.$$ Slope $1/4.2$, $\rho \approx .62$, $d_H = 5$, $d_f^2 = 15/2$. Poor tunnel; best d_H known for type and *m*.

(iv) *FFsys*, $m = 3$:

$$\begin{bmatrix} 40 & 0 & 0 & 64 \\ 0 & 40 & 0 & 70 \\ 0 & 0 & 40 & 54 \end{bmatrix}.$$ Slope $1/3.2$, $\rho \approx .68$, $d_H = 4$, $d_f^2 = 6$. Good tunnel and

best d_H for type.

4-ary Rate 1. As with binary rate $1/2$, one databit at a time creates 2 code word bits by means of two length m tapped shift registers. Generators are written as the same two left-justified octals (g_1, g_2) as before, where g_1 creates the first code word bit and g_2 the second in a branch label pair. This bit pair is mapped to 4-ary modulation values by

$$\{00, 01, 10, 11\} \longrightarrow \{W, X, Y, Z\}, \tag{4.22}$$

where W, X, Y, and Z take values in the set $\{3, 1 - 1, -3\}$ (take the second bit in a pair as the LSB). For example, $[3\text{-}1\ 1\ \text{-}3]$ means $00 \rightarrow 3, 01 \rightarrow -1, 10 \rightarrow 1, 11 \rightarrow -3$. Hereafter, this will be called the Gray map. All encoders are feed forward. The CC characteristic of most of these encoders is plotted in Figure 4.13, and the slope is listed here.

Asymptotically, the linear CC slope is the same as saying that the code symbol SER is $\approx Q(\sqrt{d_s^2 E_s/N_0})$, in which d_s^2 is the operating Euclidean square distance relative to E_s under the condition that intrinsic subtraction is applied. For 4-ary modulation, the relation between the slope and d_s^2 is derived as follows. Let the slope be determined by two points measured over an AWGN channel at SNRs $E_s^{(1)}/N_0$ and $E_s^{(2)}/N_0$. The input to the CC-BCJR is the four-dimensional LLRs of the 4-ary modulation symbols, which can be seen as an equivalent 4-ary modulator over an AWGN channel at these two SNRs. The AWGN-channel SER of such a modulator is $\approx 1.5\, Q(\sqrt{0.4E_s/N_0})$, and the code word letter SER at the CC-BCJR output is $\approx Q(\sqrt{d_s^2 E_s/N_0})$. We then have that the measured slope should be

$$\text{slope} \approx \frac{\log Q(\sqrt{0.4E_s^{(1)}/N_0}) - \log Q(\sqrt{0.4E_s^{(2)}/N_0})}{\log Q(\sqrt{d_s^2 E_s^{(1)}/N_0}) - \log Q(\sqrt{d_s^2 E_s^{(2)}/N_0})} \approx 0.4/d_s^2. \tag{4.23}$$

The operating d_s^2 is thus close to the value $0.4/\text{slope}$.

With no instrinsic subtraction ("with AP"), an operating Euclidean square distance d^2 of the decoding with respect to E_s can be measured in the same way. The code word letter SER is $\approx Q(\sqrt{d^2 E_s/N_0})$, which can equally well be written $Q(\sqrt{d^2 r_{cc} E_b/N_0})$, where $r_{cc} = 1$ now has dimensions databits/4-ary symbol. In effect, such an operating distance is found by solving the inverse Q function to find the d that leads to the observed SER. Because it includes instrinsic subtraction, we can expect d_s^2 to lie somewhat below d^2.

One can compute the classical square free distance d_f^2 by a program similar to Program 2A.2 or estimate it by $d^2 r_{cc}$ observed at a high E_b/N_0. This distance is with respect to the databit E_b, and the databit error rate is asymptotically $Q(\sqrt{d_f^2 E_b/N_0})$. It assumes that there is no intrinsic subtraction. Although d^2 applies to the code word letter error rates and d_f^2 applies to the databit rate, both are driven by the same underlying mechanism, and d_f^2 can be observed to lie near $d^2 r_{cc}$.

Listed below is the result of a code search performed for this book. Several thousand random encoder tap set and Gray map combinations are evaluated in the manner given in Section 4.3.3, wherein the convolutional code behavior in FTN iterative decoding is predicted from tests of the decoder working alone over an AWGN channel at two E_b/N_0, 1 and 5 dB, the first chosen to lie in the tunnel and the second chosen in the middle of the CC slope. The data BER at 1 dB indicates the tunnel quality (see Section 4.3.3), and the SER and BER at 5 dB indicate the quality of the overall CC characteristic. In a first pass, convolutional encoders are short-listed that have a chance of good performance. In a second pass for the short list, all 24 Gray maps are evaluated to find the best. Finally, a full test at a range of E_s/N_0 is run for good candidates in order to measure actual parameters.

Shown with each code generator are the CC slope and d_f^2, together with the probability that a start state chosen at random leads to this d_f^2 (with other starts d_f^2 is higher); d_f^2 and the probabilities are found by a minimum distance program. After these follows the observed operating d^2 near convergence (with AP) and d_s^2 (from the slope, no AP). Note that the values d_f^2, d^2, d_s^2 can be directly compared because the code rate is 1.

It was observed that good generators are not systematic and removal of AP information costs about 1.3 in d^2, compared to about 1 with binary-modulation codes.

(i) (7,5), Map [-3 3 -1 1], FF, $m = 2$. Slope 0.21, $d_f^2 = 3.6$ with prob. 1. $d^2 \approx 3.3, d_s^2 \approx 1.9$.

(ii) (54,44), Map [-3 3 -1 1], FF, $m = 3$. Slope 0.21, $d_f = 3.6$ with prob. 1. $d^2 \approx 3.1, d_s^2 \approx 1.9$.

(iii) (54,44), Map [-3 -1 3 1], FF, $m = 3$. Slope 0.21, $d_f = 3.6$ with prob. 0.5 and $d_f = 4.4$ with prob. 0.25. $d^2 \approx 3.5, d_s^2 \approx 1.9$. The new Gray map leads to better d_f but a worse tunnel.

(iv) (74,54), Map [3 -3 1 -1], FF, $m = 3$. Slope 0.16, $d_f = 4$ with prob. 0.25. $d^2 \approx 3.7, d_s^2 \approx 2.5$. Better d_f but weaker tunnel than (54,44).

(v) (52,76), Map [3 1 -3 -1], FF, $m = 4$. Slope 0.18, $d_f = 4$ with prob. 1. $d^2 \approx 3.5, d_s^2 \approx 2.2$. Better tunnel than (74,54).

(vi) (612,224), Map [3 1 -1 -3], FF, $m = 7$. Slope 0.098, $d_f = 5.6$ with prob. 0.06. $d^2 \approx 4.1, d_s^2 \approx 2.7$. Example of strong code with weak tunnel; full minimum distance available only at very high E_b/N_0.

4-ary Rate 4/3. Figure 4B.1 shows a method of rate 4/3 convolutional code generation based on the usual rate 2/3 feed forward binary encoder. Pairs of data bits drive six tapped shift registers, which produce triples of bits; for $n = 1, 2, \ldots$, the rate 2/3 structure is applied twice, to produce two triples. These six bits are mapped to three 4-ary symbols as follows: First, bits A,...,F in the two 3-tuples are mapped in some way to bits $1, \ldots, 6$ in the three 2-tuples that will serve as Gray map inputs; this is called the 3:2 map. Second, each 2-tuple maps to $\{3, 1, -1, -3\}$ in the manner defined at Eq. (4.22). The six shift registers have taps g_{ij} as defined above for binary rate 2/3 encoders. The end result is that four databits map to three 4-ary symbols.

The six tap sets and two maps need to be jointly optimized to produce the best combinations of minimum distance and tunnel. In a search performed for this book, the 3:2 maps were confined to 5 fundamentally different ones, of which the most successful were

$$\text{Map "s": A,B,...,F} \rightarrow [1, 3, 5, 2, 4, 6] \tag{4.24}$$

$$\text{Map "t": A,B,...,F} \rightarrow [1, 2, 3, 5, 6, 4] \tag{4.25}$$

$$\text{Map "n": A,B,...,F} \rightarrow [1, 2, 3, 4, 5, 6] \tag{4.26}$$

These maps have various properties; for example, map "s" preserves systematicity if it exists in the binary encoder.

As with rate 1, several thousand random encoder tap sets and Gray map combinations are evaluated over an AWGN channel at two E_b/N_0 for each 3:2 map; for good candidates all 24 Gray maps are tested. The evaluation E_b/N_0 during the code search are 2 and 6 dB. Slope and d_f^2, d^2, d_s^2 are given as with rate 1, with d_f is with respect to databit energy, but now $E_b = 3E_s/4$. As a rule longer memory encoders perform more poorly. Systematic and semisystematic encoders perform best; a search over general FF encoders usually leads to these.

It was observed that d_f is reduced at the same memory compared to $r_{cc} = 1$ codes, as one would expect, and AP removal costs roughly $(4/3) \times 0.5$ of the value of d^2.

(i) $\begin{bmatrix} 4 & 0 & 6 \\ 0 & 4 & 4 \end{bmatrix}$. FFsys, $m = 1$, 3:2 map t, Gray [1 -1 3 -3]. Slope 1.02, $d_f \approx 0.97$.

$d^2 \approx 0.72$, $d_s^2 \approx 0.39$. Good at very low SNR; otherwise a poor code.
Another good map combination is 3:2 map t, Gray [1 3 -3 -1].

(ii) $\begin{bmatrix} 4 & 0 & 7 \\ 0 & 4 & 5 \end{bmatrix}$. FFsys, $m = 2$, Map n, Gray [1 -1 3 -3]. Slope 0.87, $d_f \approx 1.6$.

$d^2 \approx 1.2$, $d_s^2 \approx .48$. Best for code type and memory.
Another combination, good at low SNR, is 3:2 map t, Gray [1 -3 -1 3].

(iii) $\begin{bmatrix} 3 & 4 & 5 \\ 4 & 3 & 7 \end{bmatrix}$. FF, $m = 2$, Map n, Gray [3 -3 -1 1]. Slope 0.34, $d_f \approx 2.7$.

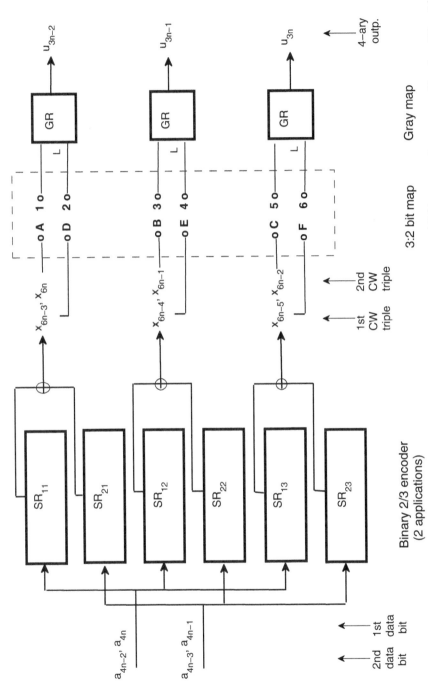

FIGURE B.1 Generation of rate 4/3 databits/4-ary symbol code words from the traditional rate 2/3 binary encoder structure. Notation: SR = shift register, GR = Gray map, CW = code word, L = least significant bit.

$d^2 \approx 1.6$, $d_s^2 \approx 1.2$. Good at moderate SNR.

Another good map combination is 3:2 map n, Gray [3 -1 -3 1].

(iv) $\begin{bmatrix} 3 & 4 & 5 \\ 4 & 3 & 7 \end{bmatrix}$. FF, $m = 2$, Map n, Gray [1 -1 3 -3]. Good at high SNR only; Cf. previous code.

(v) $\begin{bmatrix} 4 & 0 & 56 \\ 0 & 4 & 72 \end{bmatrix}$. FFsys, $m = 4$, Map n, Gray [-1 -3 1 3]. Slope 0.46, $d_f \approx 1.9$.

$d^2 \approx 1.4$, $d_s^2 \approx 0.87$. Typical longer-m encoder for low SNR.

REFERENCES

1. *J.B. Anderson, *Digital Transmission Engineering*, 2nd ed., Wiley–IEEE Press, Piscataway, NJ, 2005.

2. *J.G. Proakis, *Digital Communication*, 4th and later eds., McGraw-Hill, New York, 1995.

3. *J.B. Anderson and A. Svensson, *Coded Modulation Systems*, Kluwer-Plenum, New York, 2003.

4. *L.R. Bahl, J. Cocke, F. Jelinek and J. Raviv, Optimal decoding of linear codes for minimizing symbol error rate, *IEEE Trans. Inf. Theory*, **20**, pp. 284-287, 1974.

5. A. Lender, The duobinary technique for high speed data transmission, *IEEE Trans. Commun. Tech.*, **82**, pp. 214–218, 1963.

6. E.R. Kretzmer, Generalization of a technique for binary data transmission, *IEEE Trans. Commun. Tech.*, **14**, pp. 67–68, 1966.

7. G.D. Forney, Jr., Maximum-likelihood sequence estimation of digital sequences in the presence of intersymbol interference, *IEEE Trans. Inf. Theory*, **18**, pp. 363–378, 1972.

8. F.R. Magee, Jr., and J.G. Proakis, An estimate of the upper bound on error probability for maximum-likelihood sequence estimation on channels having a finite-duration pulse, *IEEE Trans. Inf. Theory*, **19**, pp. 699–702, 1973.

9. S.A. Fredericsson, Optimum transmitting filter in digital PAM systems with a Viterbi detector, *IEEE Trans. Inf. Theory*, **20**, pp. 479–489, 1974.

10. C. W.-C. Wong and J.B. Anderson, Optimal short impulse response channels for an MLSE receiver, *Conf. Rec.*, Int. Conf. Commun., Boston, pp. 25.3.1–25.3.5, 1979.

11. R.R. Anderson and G.J. Foschini, The minimum distance of MLSE digital data systems of limited complexity, *IEEE Trans. Inf. Theory*, **21**, pp. 544–551, 1985.

12. F.L. Vermeulen and M.E. Hellman, Reduced state Viterbi decoding for channels with intersymbol interference, *Conf. Record*, Int. Conf. Commun. Minneapolis, pp. 37B-1–37B-9, 1974.

13. G.J. Foschini, A reduced state variant of maximum likelihood sequence detection attaining optimum performance for high signal-to-noise ratios, *IEEE Trans. Inf. Theory*, **23**, pp. 605–609, 1977.

* References marked with an asterisk are recommended as supplementary reading.

14. N. Seshadri, Error performance of trellis modulation codes on channels with severe intersymbol interference, Ph.D. thesis, Dept. Electrical, Computer and System Eng., Rensselaer Poly. Inst., Troy, NY, 1986.

15. N. Seshadri and J.B. Anderson, Asymptotic error performance of modulation codes in the presence of severe intersymbol interference, *IEEE Trans. Inf. Theory*, **34**, pp. 1203–1216, 1988.

16. A. Duel-Hallen and C. Heegard, Delayed decision-feedback sequence estimation, *Proc.*, Allerton Conf. Communs., Control and Computing, Monticello, Ill., Oct. 1985; also under same title, *IEEE. Trans. Commun.*, **37**, pp. 428–436, 1989.

17. J. Mazo, Faster than Nyquist signaling, *Bell Sys. Tech. J.*, **54**, pp. 1451–1462, 1975.

18. D. Hajela, On computing the minimum distance for faster than Nyquist signaling, *IEEE Trans. Inf. Theory*, **34**, pp. 1420–1427, 1988.

19. J.E. Mazo and H.J. Landau, On the minimum distance problem for faster-than-Nyquist signaling, *IEEE Trans. Inf. Theory*, **36**, pp. 289–295, 1990.

20. K. Balachandran and J.B. Anderson, Reduced complexity sequence detection for non-minimum phase intersymbol interference channels, *IEEE Trans. Inf. Theory*, **43**, pp. 275–280, 1997.

21. A. Said, Design of optimal signals for bandwidth efficient linear coded modulation, Ph.D. Thesis, Dept. Electrical, Computer and Systems Eng., Rensselaer Poly. Inst., Troy, NY, 1994.

22. A. Said and J.B. Anderson, Bandwidth efficient coded modulation with optimized linear partial-response signals, *IEEE Trans. Inf. Theory*, **44**, pp. 701–713, 1998.

23. A.D. Liveris and C.N. Georghiades, Exploiting faster-than-Nyquist signaling, *IEEE Trans. Commun.*, **51**, pp. 1502–1511, 2003.

24. C. Douillard *et al.*, Iterative correction of intersymbol interference: turbo equalization, *Eur. Trans. Telecommun.*, **6**, pp. 507–511, 1995.

25. J.B. Anderson and J.B. Bodie, Tree encoding of speech, *IEEE Trans. Inf. Theory*, **21**, pp. 379–387, 1975.

26. D.D. Falconer and F.R. Magee, Adaptive channel memory truncation for maximum likelihood sequence estimation, *Bell Sys. Tech. J.*, **52**, pp. 1541–1562, 1973.

27. U.L. Dang, W.H. Gerstacker and S.T.M. Slock, Maximum SINR prefiltering for reduced-state trellis-based equalization, *Conf. Rec. Int. Conf. Commun.*, Kyoto, 2011.

28. F. Rusek and A. Prlja, Optimal channel shortening of MIMO and ISI channels, *IEEE Trans. Wireless Commun.*, **11**, pp. 810–818, 2012.

29. J. Hagenauer and P. Hoeher, A Viterbi algorithm with soft-decision outputs and its applications, *Proc. IEEE Global Commun. Conf.*, pp. 1680–1686, 1989.

30. M. Sikora and D.J. Costello, Jr., A new SISO algorithm with application to turbo equalization, *Proc. IEEE Int. Symp. Information Theory*, Adelaide, pp. 2031–2035, 2005.

31. A. Prlja, J.B. Anderson and F. Rusek, Receivers for faster-than-Nyquist signaling with and without turbo equalization, *Proc. IEEE Int. Symp. Information Theory*, Toronto, pp. 464–468, 2008.

32. G. Colavolpe, G. Ferrari and R. Raheli, Reduced-state BCJR type algorithms, *IEEE J. Sel. Areas Communs.*, **19**, pp. 848–859, 2001.

33. D. Fertonani, A. Barbieri and G. Colavolpe, Reduced-complexity BCJR algorithm for turbo equalization, *IEEE Trans. Commun.*, **55**, pp. 2279–2287, 2007.

34. J.B. Anderson, A. Prlja and F. Rusek, New reduced state space BCJR algorithms for the ISI channel, *Proc. IEEE Int. Symp. Information Theory*, Seoul, 2009.

35. J.B. Anderson and A. Prlja, Turbo equalization and an M-BCJR algorithm for strongly narrowband intersymbol interference, *Proc. Int. Symp. Information Theory and Its Applications*, Taichung, Taiwan, pp. 261–266, 2010.

36. A. Prlja and J.B. Anderson, Reduced-complexity receivers for strongly narrowband intersymbol interference introduced by faster-than-Nyquist signaling, *IEEE Trans. Commun.*, **60**, pp. 2591–2601, 2012.

37. A. Prlja, Reduced receivers for faster-than-Nyquist signaling and general linear channels, Ph.D. thesis, Electrical and Information Technology Dept., Lund University, Lund, Sweden, 2013.

38. J.B. Anderson, Faster than Nyquist signaling for 5G communication, in Luo and Zhang, eds, *Signal Processing for 5G: Algorithms and Implementations*, Wiley, Chichester, UK, 2016.

39. J.B. Anderson and K.E. Tepe, Properties of the tailbiting BCJR decoder, in B. Marcus and J. Rosenthal, eds, *Codes, Systems, and Graphical Models*, Springer, New York, pp. 211–238, 1999.

40. D.E. Knuth, *The Art of Computer Programing*, vol. 3, Searching and Sorting, Addison-Wesley, Reading, Mass., 1973.

41. B. Hayes, The higher arithmetic, *American Scientist*, **97**, pp. 364–368, 2009.

42. J.B. Anderson and M. Zeinali, Best rate 1/2 convolutional codes for turbo equalization with severe ISI, *Proc. IEEE Int. Symp. Information Theory*, Cambridge, Mass., pp. 2366–2370, 2012.

43. J.B. Anderson and M. Zeinali, Analysis of best high rate convolutional codes for faster than Nyquist turbo equalization, *Proc. IEEE Int. Symp. Information Theory*, Honolulu, pp. 1987–1991, 2014.

44. J.B. Anderson and S. Mohan, *Source and Channel Coding*, Kluwer, Boston, 1991.

45. R. Johannesson and K. Sh. Zigangirov, *Fundamentals of Convolutional Coding*, 2nd. ed, John Wiley, Hoboken, NJ, 2015.

46. M. McGuire and M. Sinha, Discrete-time faster-than-Nyquist signaling, *Proc. IEEE Global Commun. Conf.*, Miami, 2010.

5

MULTICARRIER FTN

INTRODUCTION

Chapter 5 extends the previous chapter to the case where a transmission has many closely spaced subcarriers, each of which can be a classical time-accelerated FTN signal. More is possible than time acceleration. The subcarriers, which in a standard system would be nominally orthogonal to each other, can be spaced much more closely in frequency. This is *frequency FTN*. Just as with pulses accelerated in time, subcarriers can be "frequency squeezed" to some degree before a distance loss occurs. There is a Mazo limit in frequency squeeze alone, and a two-dimensional limit in *both* time and frequency, above which the error performance is asymptotically that of isolated orthogonal pulses. Furthermore, the gains from these two sources are more or less separate. Whereas a bandwidth reduction of about 30% was available from time acceleration alone, as much as 50% is available in this chapter.

Frequency FTN is more challenging than time FTN. Subcarriers have a phase relationship, and it affects signal distance. As if time and frequency spacing were not variables enough, there are many ways to improve distance with the phase variable. Being carrier transmission, the signals have I and Q components. Detection is more difficult because frequency cochannel interference (CCI) needs equalization in addition to time ISI. When added to codeword detection, this can lead to a three-dimensional BCJR receiver.

The subcarrier FTN problem began with References 3 and 4, and that line is followed here. The subject has seen less development than time FTN, but there are

Bandwidth Efficient Coding, First Edition. John B. Anderson.
© 2017 by The Institute of Electrical and Electronics Engineers, Inc. Published 2017 by John Wiley & Sons, Inc.

software and hardware receivers to report. There are also other lines of research, which are not our classical approach, but which hint at the gains available when one gives up orthogonal subcarriers. There are also many similarities to magnetic tape recording and particularly to OFDM (orthogonal frequency division multiplex), important technologies in their own right. One can view frequency-squeezed FTN as an enhanced OFDM; for an introduction to OFDM we can recommend [1,2].

Section 5.1 presents classical frequency FTN, comments on capacity, and defines a spacing variable ϕ, the frequency analog to the acceleration parameter τ. The complex subject of minimum distance is Section 5.2. The section suggests good signal configurations and it estimates the two-dimensional Mazo limit. Section 5.3 recounts some receiver test results and presents some cousins to the classical scheme.

5.1 CLASSICAL MULTICARRIER FTN

A good place to start is the "$2WT$" result of Theorem 3.2: Signals with bandwidth W running for T seconds can sustain about $2WT$ orthogonal dimensions. The original theorem required that these signals be made of sinc(t/T) pulses, which exist only in the limit of physical pulses, but they provide a useful framework for understanding time–frequency FTN. A way to view the signal set is Figure 5.1. It is a framework, with a sinc pulse associated with each point in a $K \times N$ point array. The rows represent subcarriers at $f_k = k/T$ Hz, $k = 0, \ldots, K - 1$, and the columns are the centers of sinc pulses at nT seconds, $n = 0, \ldots, N - 1$. No matter what T is, the area of the array is $(K/T)NT = KN$ Hz-s, half the available orthogonal dimensions. The number in fact does not depend on T, and we will often let T be 1. However, the value of T gives the physical frequencies f_k and the times τT.

The array will serve as a reference for time–frequency FTN systems. As before, we will base the signals on linear modulation with a unit-energy base pulse $h(t)$. Now the signal is made up of subcarriers, each of which can be a time-FTN signal in the "accelerated" formulation Eq. (2.50); the outcome is

$$s(t) = \sqrt{2E_s} \sum_{k=0}^{K-1} \sum_{n=0}^{N-1} u_{k,n}^I h(t - n\tau T) \cos 2\pi(f_c + f_k)t \quad -$$
$$u_{k,n}^Q h(t - n\tau T) \sin 2\pi(f_c + f_k)t. \tag{5.1}$$

There are now in-phase (I) and quadrature (Q) PAM modulator symbols u^I and u^Q, and E_s is the average symbol energy of each. A carrier is required and it is f_c Hz.[1] When $h(t)$ is a sinc pulse and $\tau = 1$, the time–bandwidth occupied is $[-T/2, NT - 1/2] \times [f_c - 1/2T, f_c + (K - 1/2)/T]$ Hz-s, counting positive frequencies. When $h(t)$ is a general T-orthogonal pulse and $\tau = 1$, the time and bandwidth are somewhat larger—but the physical-world product, divided by KN, still tends to 1 as $K, N \to \infty$.

---

[1] Another common notation uses complex numbers. The signal is the real part of $\sqrt{2E_s/T} \sum_k \sum_n u_{k,n} h(t - n\tau T) \exp(j2\pi(f_c + f_k)t)$, in which u is the complex symbol $u^I + ju^Q$.

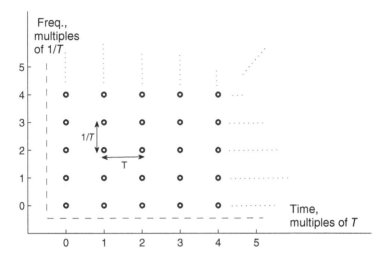

FIGURE 5.1   Reference framework for time–frequency FTN, based on sinc-pulse signaling. Under FTN signaling, the spacing will become τT and ϕ/T.

In time-only FTN, we have $\tau < 1$ and the time occupancy reduces asymptotically in N to τNT. In frequency FTN, we can define a *frequency squeeze factor* ϕ to be the subcarrier spacing ϕ/T in terms of $1/T$ Hz. The frequency occupancy is now $\phi K/T$, asymptotically in K. A new reference framework in the style of Figure 5.1 represents the signals, with horizontal point spacing τT and vertical spacing ϕ/T. The symbol density of the signal becomes altogether $2/\phi\tau$ symbols/Hz-s.[2] The factor "2" here arises because of the independent I and Q signals.

A difficulty remains, namely, that the subcarriers are unlikely to be strictly orthogonal when $h(t)$ is not a sinc, even when $\phi \geq 1$. The problem is that $\int H(f)H'(f)\,df$ is not necessarily zero when $H(f)$ and $H'(f)$ are transforms of pulses at parallel times but on subcarriers spaced $1/T$ Hz; that is, the subcarrier signals are not orthogonal. For example, the frequency dual of the spectral antisymmetry condition (Property 1.1) asserts that a sufficient condition would be that $h(t)$ is antisymmetric about the point $(h(0)/2, T)$. This is inconvenient, and it is easily verified that favorite pulses fail strict orthogonality. However, reasonable T-orthogonal time pulses are very nearly orthogonal when they appear on subcarriers separated $1/T$ Hz, enough so that the error can be ignored. Whether it is or not, CCI will become a definite feature when frequency squeezing is applied.

Incorporating ϕ and τ and rewriting Eq. (5.1) into a form with a single I and Q, we get

$$\sqrt{2E_s}\left[I(t)\cos 2\pi f_c - Q(t)\sin 2\pi f_c\right],$$

---

[2]An older name in physics for this "tiling" of time–frequency is Weil-Heisenberg system. An early exploration from a communication point of view is Reference 11.

where

$$I(t) = \sum_{k=0}^{K-1} \sum_{n=0}^{N-1} \left[u_{k,n}^I h(t - n\tau T) \cos 2\pi k\phi t/T - u_{k,n}^Q h(t - n\tau T) \sin 2\pi k\phi t/T \right] \text{ and}$$

$$Q(t) = \sum_{k=0}^{K-1} \sum_{n=0}^{N-1} \left[u_{k,n}^Q h(t - n\tau T) \cos 2\pi k\phi t/T + u_{k,n}^I h(t - n\tau T) \sin 2\pi k\phi t/T \right].$$

$$(5.2)$$

The signals $\cos 2\pi k\phi t/T$ and $\sin 2\pi k\phi t/T$ can be thought of as the quadrature subcarriers, although they do not exist as physical sine waves. The normalized square Euclidean distance between two signals $s^{(1)}(t)$ and $s^{(2)}(t)$ is $d^2 = (1/2E_b) \int |s^{(1)}(t) - s^{(2)}(t)|^2 dt$. In the limit of f_c only the difference in I and Q matters, and that depends only on the differences Δu^I and Δu^Q. The normalized square distance in the binary case may then be written[3]

$$(1/2) \int \left[|\Delta I(t)|^2 + |\Delta Q(t)|^2 \right] dt, \qquad (5.3)$$

$\Delta I(t)$ and $\Delta Q(t)$ are defined just as I and Q in Eq. (5.2) but with $\Delta u_{k,n}^I$ and $\Delta u_{k,n}^Q$ instead of $u_{k,n}^I$ and $u_{k,n}^Q$.

Some Multicarrier Signals. Because of the subcarrier structure, it is challenging to visualize the I and Q components of a time-frequency FTN signal. The next three figures show an example of the signals in a binary 3-subcarrier system. The three in-phase signals are I_0, I_1, and I_2, the quadrature signals are Q_0, Q_1, Q_2, and the totals of each in Eq. (5.2) are I_{tot} and Q_{tot}. When quadrature modulated at carrier f_c Hz, these represent the signals from all six u streams. For clarity, all the u^Q symbols are zero, and the u^I symbols are zero except for $u_{k,0}^I = 1$ and $u_{k,5}^I = 1$ on all three subcarriers. This makes it clear what subcarrier transmission does to each symbol. The x-axis is in symbol times.

Figure 5.2 shows the case when $\phi = 0.8$. Even though all u^Q are 0, both I and Q responses appear everywhere except Q_0. Observe that when time advances $\ell\tau T$, the sin and cos phase advances $2\pi k\ell\phi\tau$, so that if the pulses are narrow enough, the responses to two pulses $\ell\tau T$ apart will be almost identical when $k\ell\phi\tau$ is an integer for each k in use. This is evident in Figure 5.2, where $5 \times \phi\tau$ is 4. The ripple in I and Q grows in frequency with the subcarrier number k.

Figure 5.3 shows the same, but now $\phi = 0.3$ with τ still 1. The ripples are slower, reflecting the smaller subcarrier frequencies, and because $5\phi\tau = 1.5$ is an odd multiple of $1/2$, the pulses in I_1 and in Q_1 have opposing symmetry and in I_2 and Q_2 they are the same. Figure 5.4 repeats Figure 5.3, but with $\phi = 0.6$ and $\tau = 0.5$. The $\tau = 0.5$ doubles the width of all time responses, but once again $5\phi\tau = 1.5$ so that the

---

[3]When u is M-ary, the factor in front is $\log_2 M/2$. This chapter treats only the binary case.

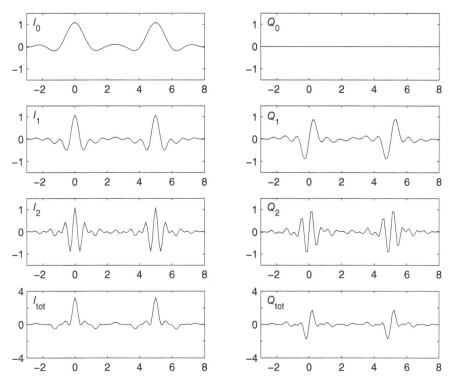

FIGURE 5.2 I and Q signals in time–frequency FTN with three subcarriers (I), $\phi = 0.8$, and $\tau = 1$. $u_{0,0}^I, u_{0,5}^I, u_{1,0}^I u_{1,5}^I u_{2,0}^I u_{2,5}^I$ are 1; all other symbols are 0.

same symmetries hold as in Figure 5.3. The wider pulses damage the symmetries but they are still evident.

All the Q symbols here were 0. In a full transmission, all the u^I and u^Q symbols would take ± 1 values and the Q symbols would lead to their own similar behavior. The composite I and Q would not be easily disentangled by eye.

OFDM; Slepian's Problem. OFDM, a method of stacking subcarriers that is similar in form to frequency FTN, avoids the not-quite-orthogonal problem by means of a fast Fourier transform that produces I and Q. Frequency-alone FTN can be viewed as an extension of OFDM to subcarriers that are significantly nonorthogonal. One can also employ pulses that are not orthogonal for any T. Such pulses can be solutions to Slepian's problem, which asks what pulse minimizes time–bandwidth for a fixed-length message when a given fraction of the pulse energy is allowed to escape the $\mathcal{W} \times \mathcal{T}$ box. These pulses are taken up in Chapter 7.

The Shannon Limit. A useful Shannon limit is less complicated to estimate for subcarrier FTN than it is for time FTN. The subcarriers more or less fill a time--bandwidth block of size $\phi K/T$ Hz and τNT seconds with uniform power, and

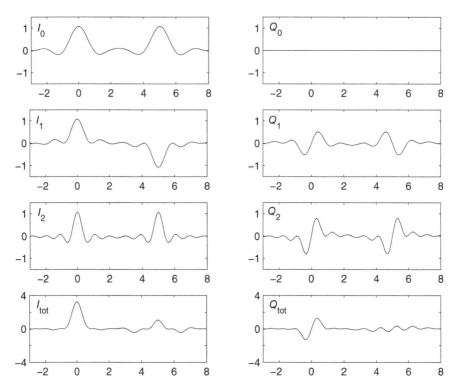

FIGURE 5.3 I and Q signals in time–frequency FTN with three subcarriers (II). As Figure 5.2 but $\phi = 0.3, \tau = 1$.

spectral side lobes play little role when K and N are not small. The Gaussian-noise capacity of the block is approximately C_{sq} in Section 3.2 per Hz-s. From the discussion there and Eq. (3.6), this is $\log_2(1 + 2E_{cu}/N_0)$ bits/Hz-s, where E_{cu} is the available energy per channel use (dimension) in the capacity calculation (note that Eq. (3.6) applies at both passband and baseband).

With these assumptions in place, we can calculate E_{cu} and the transmission rate in bits/Hz-s implied by the coding and modulation parameters. The energy expended by the modulator is $2NKE_s$ J per block, and there are about $2\phi\tau KN$ channel uses in the block. Therefore

$$E_{cu} = \frac{2KNE_s}{2\phi\tau KN} = E_s/\phi\tau = E_b r_{cc}/\phi\tau, \quad \text{or} \tag{5.4}$$

$$E_b = \phi\tau E_{cu}/r_{cc} = E_s/r_{cc}, \tag{5.5}$$

where E_b as always is the databit energy. Here for future use is the coding rate r_{cc} from Chapter 4 applied to the modulator, in databits/M-ary modulator symbol; if there is no coding, $r_{cc} = \log_2 M$.

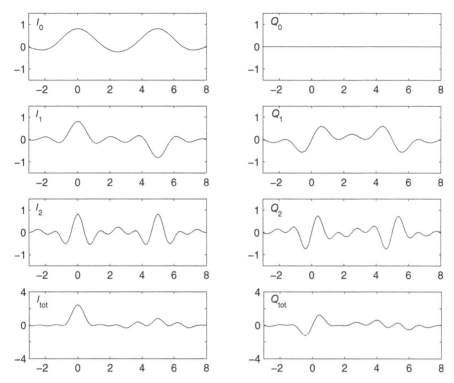

FIGURE 5.4 I and Q signals in time–frequency FTN with three subcarriers (III). As Figure 5.2 but $\phi = 0.6, \tau = 0.5$.

To find the Shannon limit, we require the rate at which the system works, in bits/Hz-s. Find first that the databits per block are $2KNr_{cc}$. Then the capacity per Hz and second over the block needs to be

$$\frac{2KNr_{cc} \text{ bits}}{\phi\tau KN \text{ Hz-s}} = 2r_{cc}/\phi\tau \quad \text{bits/Hz-s.} \tag{5.6}$$

This is the time-only FTN bit density in Eq. (4.2), with an added factor $1/\phi$ for the frequency squeeze. The last step is to apply the procedures in Section 3.2 and Appendix 3A to calculate the Shannon limit relation between BER and the databit E_b/N_0. The needed capacity at BER=0 is $2r_{cc}/\phi\tau$ bits/Hz-s and Program 3A.4 computes the relation, with the input rate set to $2r_{cc}/\phi\tau$.

Example 5.1 *(Shannon Limit When $\phi\tau/r_{cc} = 0.5$)*
In Section 5.2, it will develop that the Mazo limit may lie as low as $\phi\tau = 0.5$ for uncoded time–frequency FTN. This value, which corresponds to 4 bits/Hz-s, allows some interesting comparisons. Applying Program 3A.4 with rate = 0.5, one

estimates the Shannon limit to lie at $E_b/N_0 \approx 5.7$ dB when the databit BER $\approx 10^{-5}$. The BER of the physical FTN system at the Mazo limit is $\approx AQ(\sqrt{2E_b/N_0})$; taking $A \approx 2$ yields BER 10^{-5} at about 9.8 dB, about 4 dB from the limit. Simple 4PAM modulation works in the same 4 bits/Hz-s and requires 13.7 dB. Thus the uncoded FTN system gains about 4 dB in energy over 4PAM. A comparison can be made to Figure 4.15, which shows time-only uncoded FTN with $\tau = 1/2$ and the same 4 bits/Hz-s; this gains 2 dB over 4PAM. The added complexity of time–frequency FTN thus gains an additional 2 dB.

A much more difficult capacity calculation is to include the details of the signaling, rather than just the overall PSD shape as we have done here. A start on that calculation appears in Reference 22.

5.2 DISTANCES

Because there are I and Q symbols, dimensions of time and frequency, and phase off-sets to contend with, time–frequency FTN distances are a challenge. While distances-–hence asymptotic error performance—can be found, there are a great many design combinations, and it is difficult to know that the best one has been discovered. In what follows, we first set up the distance problem and then find optimal signal minimum distances and Mazo limits under certain design strategies. These are a guarantee that signal sets at least this good exist, but better ones may yet be discovered.

5.2.1 Finding Distances

In principle, normalized square distance between two signals is given by Eq. (5.3), in which $I(t)$ and $Q(t)$ are given by Eq. (5.2). The symbols u in Eq. (5.2) are replaced wherever they appear by *symbol differences* Δu, and these are of two types: Δu^I and Δu^Q. The set of differences defines an error difference event in the style of Section 2.5.1, but now I and Q and subcarriers as well as the event's start time play a role. In what follows, we describe events by two $K' \times N'$ matrices, $\boldsymbol{\Delta}^I$ for the in-phase symbol differences and $\boldsymbol{\Delta}^Q$ for the quadrature differences. Symbols over K' subcarriers and N' intervals affect the event and all other differences are zero.

As an example, consider $\phi = 0.8$, $\tau = 0.7$ and a 30% root RC pulse, with error event differences

$$\boldsymbol{\Delta}^I = \begin{bmatrix} 2 & -2 & 0 \\ -2 & 2 & 0 \end{bmatrix}, \qquad \boldsymbol{\Delta}^Q = \begin{bmatrix} 0 & 2 & -2 \\ 0 & 2 & -2 \end{bmatrix}. \tag{5.7}$$

If all subcarriers start from phase 0 and time 0, the outcome is normalized square distance 6.96. However, one soon notices that the outcome depends on the start time of the event. For example, starting at time 1.25 symbol intervals leads to distance almost 20, and starting at 0.34 yields only ≈ 1.13. The underlying problem here is

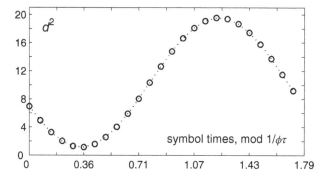

FIGURE 5.5 Square distance versus start time in symbol times modulo $1/\phi\tau$ for event Eq. (5.7). Parameters are $\phi = 0.8, \tau = 0.7$. Distance can take only 25 values.

that the symbols are not necessarily synchronized with the subcarrier phases, and there is a global best and worst time for an error event to start.

The problem is shown in more detail in Figure 5.5. The x-axis is the event start in symbol times. The worst case square distance occurs at ≈ 0.34 and a pattern repeats each $1/\phi\tau$ symbols. The pattern occurs because from Eq. (5.2) a time delay of $t_o = \ell T/\phi$, ℓ an integer, will lead to the same sin and cos phases modulo 2π. In accelerated symbol times (see Eq. (2.50)), this is $t_o/\tau T = 1/\phi\tau$ symbols, taking the smallest ℓ. In the example, $1/\phi\tau = 1/.56 \approx 1.79$. When only two subcarriers contribute to an error event, the pattern is in fact a sine for any ϕ and τ, which means that the entire pattern can be constructed from three points.[4]

When there is no control of phase offsets among the subcarriers, any distance in the figure can occur, and the worst case sets the receiver asymptotic error rate. Although that case is unlikely, it is clear that the scenario should be avoided. When $\phi\tau = i/j$, i and j positive integers without a common factor, the system returns to its time-0 phase and the distances that occur are only those at multiples of $(1/\phi\tau)/j$. This is called *synchronous* time–frequency FTN. Now the worst-case distance is the minimum of only the j allowed distances. In Figure 5.5, $\phi\tau = 0.56 = 14/25$ and there are 25 points (shown by circles, when phase is 0 at time 0). The picture repeats in 25 steps of 1/14 symbol.

If i and j are small, it should be possible to avoid a poor global minimum. Some experiment shows that $j = 2$ or 3 are good choices.

At least two other phase and delay tactics can be applied. First, a pattern of *pulse delays* δ_k can be implemented across the subcarrier streams. Each can be delayed in a step pattern, for example, $\delta_k = 0.1k$ for stream $k = 0, 1, 2, \ldots$; or delays can cycle, for example, $0, 0.5, 0, 0.5, \ldots$. Second, *subcarrier delays* can be applied, for example, phase steps of size $2\pi k\epsilon$. Combinations of all of these can be applied. To a large degree, they accomplish the same aim, to avoid placing the strong part of a pulse at a weak spot in a subcarrier.

---

[4] A proof of the sine shape by F. Rusek appears in Reference 4.

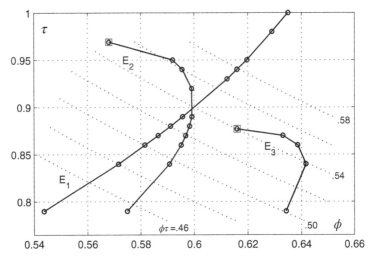

FIGURE 5.6 Trajectories in τ–ϕ space for which $d^2_{\min} = 2$, for the three event families in Eq. (5.8). 30% root RC time–frequency FTN. Dotted lines show contours of constant $\phi\tau$ product.

5.2.2 Minimum Distances and the Mazo Limit

We turn now to finding the minimum-distant error event. Because there are I and Q difference symbols, the search effort is the square of time-only FTN. As well, there are the many FTN variants to consider. Finding a good system is thus a complex undertaking, but a number of simplifications also exist.

The end result of this subsection is estimation of the binary Mazo limit, that is, the location of $\phi\tau$ products leading to square distance 2, and in particular, the smallest such product. Nonbinary systems and systems with $d^2_{\min} < 2$ are promising subjects for future work. Some individual tests of the latter appear later in Section 5.3.

The Mazo limit is now a boundary in ϕ–τ space. Figures 5.6 and 5.7 are both plots over this space. Figure 5.6 shows the $d^2_{\min} = 2$ achieving trajectories of the three individual error events

$$\mathcal{E}_1 : \quad \mathbf{\Delta}^I = \begin{bmatrix} -2 & 0 \\ 2 & -2 \\ 0 & -2 \end{bmatrix}, \qquad \mathbf{\Delta}^Q = \begin{bmatrix} 0 & -2 \\ 2 & -2 \\ -2 & 0 \end{bmatrix},$$

$$\mathcal{E}_2 : \quad \mathbf{\Delta}^I = \begin{bmatrix} -2 & 0 & 2 \\ 2 & -2 & 0 \end{bmatrix}, \qquad \mathbf{\Delta}^Q = \begin{bmatrix} 0 & -2 & 0 \\ 0 & -2 & 2 \end{bmatrix},$$

$$\mathcal{E}_3 : \quad \mathbf{\Delta}^I = \begin{bmatrix} -2 & 2 & 0 \\ 2 & -2 & 0 \end{bmatrix}, \qquad \mathbf{\Delta}^Q = \begin{bmatrix} 0 & -2 & 2 \\ 0 & -2 & 2 \end{bmatrix}. \qquad (5.8)$$

Each is obtained by finding the subcarrier spacing ϕ yielding $d^2_{\min} = 2$ for each of a set of time accelerations τ. The base pulse is 30% root RC. Interesting properties can be observed in the calculations.

- Many difference error events lead to an identical outcome, because of symmetries in time and phase. The set of these is called an error family. The events $\mathcal{E}_1, \mathcal{E}_2, \mathcal{E}_3$ are each the "head" of a family. It is only necessary to calculate for one such head.

- For each event family, there can exist a ϕ value below which no τ leads to $d_{\min}^2 = 2$, or a τ below which no ϕ leads to 2; the subcarrier spacing or time acceleration alone precludes 2. These points are indicated by boxes in Figure 5.6. For an error event with box coordinates (τ_o, ϕ_o), no $\tau > \tau_o$ allows 2 when $\phi < \phi_o$, or the transposed statement, or both. There is a $d_{\min}^2$ associated with these (ϕ, τ), but it is less than 2.

- When sufficiently many families have been plotted, the Mazo limit points are the largest ϕ for a given τ or the largest τ for a given ϕ, unless these conflict, in which case the point with the largest product is taken.

- Usually, but not always, (τ_o, ϕ_o) that achieves $d_{\min}^2 = 2$ implies that ϕ_o and $\tau > \tau_o$ satisfy $d_{\min}^2 \geq 2$, and similarly for τ_o and $\phi > \phi_o$.

Underlying these comments is the fact that there is a kind of distance continuity from point to point in the plane: The $d_{\min}$-causing event at a (τ, ϕ) point will lead to nearly the same $d_{\min}$ in a region around the point. The distance of other events will change some, but not fall below this $d_{\min}$. With enough displacement of (ϕ, τ), a new event leads to a new $d_{\min}$. Finding $d_{\min}$ can be viewed as tracking a small number of critical events.

Plotting the Mazo limit is thus a matter of tracking a few event family heads. Most often the families are those that play a role with other FTN parameter values as well. One can track through points in the plane a family thought to lead to the limit, and occasionally verify that assumption with an extensive search over many times and frequencies; if the present family no longer leads to $d_{\min}$ a new family has been discovered. The final result will be a good estimate of the limit location, even if the precise (τ, ϕ) are not found. Some advanced distance-finding algorithms appear in Reference 4. Events that extend over many symbols and subcarriers have a smaller multiplicity factor and make progressively less contribution to the receiver error rate. The total effect of such extreme events can only be measured by receiver tests.

Figure 5.7 shows the approximate Mazo limit location for uncoded time–frequency FTN based on 10% and 30% root RC pulses. The dashed lines represent nonsynchronous systems with no imposed pattern of phase or time offsets; the distances are the worst case of curves like Figure 5.5. Typically three to five event families determine the whole Mazo line. The figure was confirmed by events out to 4×7 and 7×4 subcarriers $\times$ symbols, but the contributing critical events were smaller, typically 3×3, 2×4, or 4×2. The solid curve represents 10% root RC for which the subcarrier pulse trains are delayed $0, 0.5, 0, 0.5, 0, \ldots$ symbol intervals.

The 30% root RC pulse achieves $\phi\tau$ nonsynchronous products as low as 0.6, with $\phi \approx 0.67$ and $\tau \approx 0.88$. This can be compared to the time-only Mazo limit, which lies at $\tau = 0.703$ and the frequency-only limit, which is ≥ 0.64. The 10% pulse achieves $\phi\tau = 0.55$, and nearly 0.53 with the delay pattern; such patterns improve the 30%

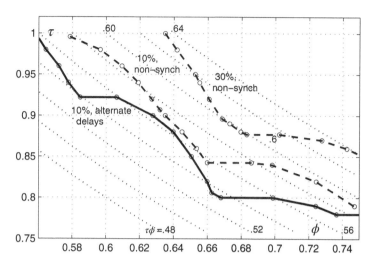

FIGURE 5.7 Three estimated Mazo limit positions in $\phi-\tau$ space for uncoded FTN: 30% and 10% root RC pulse nonsynchronous (dashed), and 10% FTN with pulse trains delayed $0, 0.5, 0, 0.5, , \ldots$ (solid line). (Data taken from Reference 4 and private sources.)

case similarly. Synchronous 10% systems have been reported with $\phi\tau$ as low as 1/2. This is a doubling in bits/Hz-s compared to OFDM, without loss of error performance.

5.3   ALTERNATIVE METHODS AND IMPLEMENTATIONS

Receiver performance is not as well researched for classical time–frequency FTN as the time-only case, but there are nonetheless interesting methods to report and even working chips. The design principles are the same: An iterative design with several processors is often required, and if convergence occurs, coded FTN achieves the convolutional CC line defined in Section 4.1.2. The CC line is reached even when (ϕ, τ) lie far below the Mazo limit, but convergence occurs at progressively higher E_b/N_0 as the $\phi\tau$ product falls. Most of the methods in this section use successive interference cancellation (SIC), which was introduced in Section 2.2.2. All are binary and the majority are not coded, in which case they usually seek to reach the above-Mazo-limit BER $Q(\sqrt{2E_b/N_0})$.

Receiver design becomes much more challenging when the $\phi\tau$ product is small. Practical designs have assumed that either ϕ or τ are in the range 0.9–1 so that respectively CCI or ISI can almost be ignored. When both are smaller than 0.8 or so, interference removal becomes strongly two dimensional and much more difficult, and designs for coded FTN must iterate in some sense among three or more processors. For example, the first implementation presented below includes a time-BCJR SIC, and a convolutional decoder BCJR. The two-dimensional interference problem resembles two-dimensional equalization for magnetic recording, and this field provides some

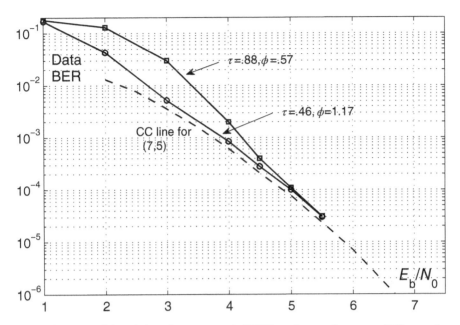

FIGURE 5.8   SIC-based time–frequency coded FTN receiver performance. (7,5) convolutional coding. 1000 blocks, length 10k. 30% root RC pulse. (Data adapted from Reference 5.)

inspiration. Here too one would like to employ an added coding dimension. Some relevant papers in magnetic recording are References 6–9.

An ISI-BCJR – FFT Combination. Figure 5.8 shows the performance of a receiver devised by Rusek [4,5] for coded time–frequency FTN. Two (ϕ, τ) combinations are shown, $(1.17, 0.46)$, which has product 0.54 and 1.85 bits/Hz-s, and $(0.568, 0.88)$, which has product 0.5 and 2 bits/Hz-s, both with no time or phase offsets or subcarrier synchrony. The iterative receiver is an outgrowth from [10] and has several features that have not appeared in the book. It breaks the modulation symbols in a given set of iterations into two groups, one, $\mathcal{U}_{dec}$, that it is estimating, and $\mathcal{U}_{int}$ that it treats as interference. A natural strategy is to focus the attack on whichever of the ISI or ICI is strongest, and to treat the other as noise. Thus, if ICI from neighboring subcarriers is relatively minor, $\mathcal{U}_{dec}$ includes only symbols along one subcarrier. A soft estimate of the interference from the $\mathcal{U}_{int}$ symbols is subtracted from the received signal, likelihoods of the time symbols in $\mathcal{U}_{dec}$ are estimated, and the entire procedure is repeated for each subcarrier. A second new feature is that this last estimate is performed by an ISI-BCJR that operates with colored noise (see Section 2.2.2). The colored noise comes about, among other reasons, because there are filters matched to nonorthogonal modulation pulses $h(t)$, which produce colored noise. The ISI-BCJR is a limited-trellis model with memory 5 (Section 4.2.1).

The outcome of this symbol estimation feeds the usual CC-BCJR, which decodes the (7,5) rate 1/2 convolutional code. The receiver limits itself to seven global iterations in all cases, and its error performance stabilizes at about the values shown for 7 or more subcarriers, a feature not present in some of the methods that follow. This performance is comparable to the time-only BER shown in Figure 4.14. The figure's parameter combinations lie below the Mazo limit in Figure 5.6, yet the performance reaches the CC line (were they above the limit it would mean in theory that only 1–2 ISI-BCJR iterations were needed). Rusek shows that CC line performance can actually be reached with more extreme (ϕ, τ), as low as $\phi\tau = 0.45$ with the (7,5) coding and 0.43 with (74,54). Future research may show that even lower products may be reached with more complex detection.[5] The Shannon limit for these bit densities can be estimated from Figure 3.4, the dashed "square PSD" curves; at BER 10^{-4} the plots are 2–3 dB from the limit. We can conclude from the work that attractive FTN schemes are available when the communication scenario favors stacked subcarriers, but high-performance receivers are not yet explored.

An ISI-BCJR–FFT Combination and an FTN Chip. Figure 5.10 shows coded FTN results from the Ph.D. thesis of Dasalukunte [12,13]. The receiver is again SIC-based, but it contains several innovations. The design follows the lines of OFDM as much as possible: The transmitter uses an inverse fast Fourier transform (IFFT) and the receiver an FFT; the work primarily uses the usual ϕ in OFDM, namely $\phi = 1$, together with $\tau < 1$; the time-FTN causes ISI in the receiver FFT streams, which are SIC-equalized. The FTN pulse $h(t)$ is not a root RC, but is instead the IOTA pulse[6] that is preferred in OFDM. The object is to explore FTN as an enhancement for OFDM.

Another advance is construction and test of a receiver chip in 65 nm CMOS, apparently the first FTN chip to be implemented. The IOTA pulse implementation and the many details of the chip development are traced in the thesis and a following book [12]. Figure 5.9 is a die photograph of the receiver chip. Figure 5.10 shows the measured chip performance when $\tau = 1$ (traditional OFDM) and 0.6. The chip supply voltage ranges over 0.7–1.2 V, depending on the desired clock speed. At the maximum 100 MHz speed, the databit throughput is 1 Mb/s with power consumption 9.6 mW.

We turn next to methods that have developed in parallel with classical time–frequency FTN.

The SEFDM Method. The spectrally efficient frequency-division multiplex (SEFDM) family of methods was introduced in its present form by Darwazeh and coworkers in 2009 [17]. Some further papers are References 18–20 and a hardware

---

[5]But threshold likely occurs at higher E_b/N_0 with smaller products; this is visible in Figure 5.8.

[6]IOTA = isotropic orthogonal transform algorithm. See Chapter 7. The pulse is a compromise between the Gaussian pulse and orthogonality, which achieves the time and frequency compactness of a nearly Gauss shape in an orthogonal form. See Reference 14–16.

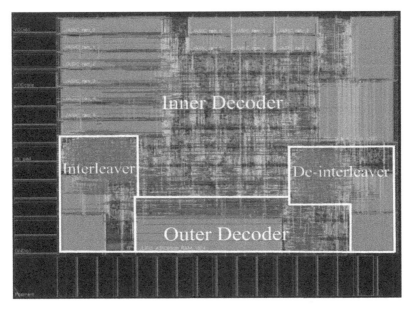

FIGURE 5.9 An FTN chip layout, fabricated by Dasalukunte. (Reproduced by permission of D. Dasalukunte, Intel Corp.)

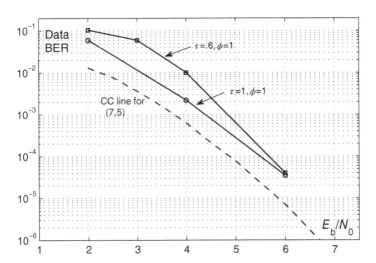

FIGURE 5.10 An FTN chip: SIC-based coded FTN receiver performance for FTN based on OFDM. (7,5) convolutional coding. IOTA pulse. (Data adapted from Reference 12.)

paper is Reference 21. These papers extend the FFT and SIC ideas in one way or another to form the receiver, and they are 4QAM frequency-only FTN. Reference 17 is a Gram-Schmidt–sphere detector design, without coding; BER performance at the theoretical $Q(\sqrt{2E_b/N_0})$ is demonstrated with frequency squeeze down to

$\phi = 0.7$. Reference 19 is an FFT–SIC design with full performance to $\phi = 0.67$; a full exposition of the FFT method is given. Reference 20 explores FFT–sphere decoder and FFT–ISI-BCJR turbo designs; the second is tested over a frequency- and time-dispersive channel.

The GFDM Method. Generalized frequency division multiplexing (GFDM) was proposed by Fettweis et al. [23]. This method does not make explicit use of subcarrier squeezing, but the transmission base pulse is mildly nonorthogonal so that some of the receiver techniques of FTN are needed.

Frequency-Domain Methods. Equalization of interference is performed in the time domain in this book, but interesting methods are available that perform it in the frequency domain, especially when subcarrier transmission is employed. A recent paper that applies such methods to FTN, coupled with an M-BCJR-like approach, is Reference 24. The paper also treats compensation of time and frequency dispersion in the channel.

LDPC and Two Staggered Signal Sets. Barbieri, Fertonani, and Colavolpe in Reference 22 propose a time–frequency FTN system, where the $\phi\tau$-compressed reference framework of pulses in the style of Figure 5.1 is split into two staggered lattices. Each separately represents either an array of orthogonal signals or at least ones that are more nearly orthogonal. Each is treated as interference to the other, and an iterative soft SIC scheme removes the interference. There is no explicit FFT.

The system is tested only with binary symbols and at a low 1–2 bits/Hz-s, but there are a number of innovations. A Gauss pulse is evaluated, since this shape simultaneously minimizes time and frequency interference, even though it is not time-orthogonal (more about the Gauss pulse in Chapter 7). The setup is explicitly 4QAM, although specifically quadrature issues are not addressed. Finally, coded FTN tests are performed with LDPC (low-density parity check) codes, a popular method that approaches closely to the Shannon limit when employed with 2PAM or 4QAM modulation, but which has significant complexity. LDPC replaces the convolutional coding that dominates earlier in the book, and there is good reason to suspect that it will perform better. It is, of course, interesting what happens when LDPC is combined with FTN whose $\phi\tau$ is significantly less than 1. At LDPC rate 1/2, essentially orthogonal transmission and 1 bit/Hz-s, the authors show BERs that lie only 1.2 dB from the Shannon limit[7] at BER 10^{-5}. When $\phi\tau$ drops to around 0.8, performance is 2–3 dB from the limit.

The question remains open how much LDPC codes need redesign for FTN, or for that matter, large-alphabet modulation. Just as convolutional codes need redesign, so also will LDPC codes. A group at EUROSAT in Paris found evidence in 2013 that major gains are available [25,26]. A recent paper with many references that

---

[7]The traditional Nyquist capacity C_{sq} in Section 3.2 is employed here but C_{sq} and C_{PSD} are very close at this low bit density. The relevant limit value at low BER is the point $(Rate, E_b/N_0) = (1, 1)$ in Figure 3.2.

focuses on the combination of nonbinary LDPC with 16QAM and 64QAM simple modulation is Zhu et al. [27]. Schemes are reported in the 3–4 bits/Hz-s range that lie a few dB from the Nyquist–Shannon limit.

5.4 CONCLUSIONS

In a dual method to time FTN, subcarriers can be squeezed together in frequency, with a squeeze factor ϕ.

Finding distances is more challenging because of I and Q signals and phase relations among the subcarriers.

The Mazo limit becomes two-dimensional in τ and ϕ. As much as twice the bits/Hz-s can be achieved without error rate loss.

Many frequency squeeze variants have been proposed. Many are related to OFDM.

REFERENCES

1. *E. Dahlman, S. Parkvall, and J. Sköld, *4G: LTE/LTE-Advanced for Mobile Broadband*, 2nd ed. Academic Press/Elsevier, Oxford, UK, 2011.

2. *A.F. Molisch, *Wireless Communications*, Wiley, Chichester, UK, 2005.

3. F. Rusek and J.B. Anderson, The two dimensional Mazo limit, *Proc. IEEE Int. Symp. Information Theory*, Adelaide, pp. 970–974, 2005.

4. F. Rusek and J.B. Anderson, Multistream faster than Nyquist signaling, *IEEE Trans. Commun.*, **57**, pp. 1329–1340, 2009.

5. F. Rusek, Partial response and faster-than-Nyquist signaling, Ph.D. thesis, Electrical and Information Technology Dept., Lund University, Lund, Sweden, 2007.

6. Y. Wu et al., Iterative detection of and decoding for separable two-dimensional intersymbol interference, *IEEE Trans. Magn.*, **39**, pp. 2115–2120, 2003.

7. J.B. Soriaga, H.D. Pfister, and P.H. Siegel, On achievable rates of multistage decoding on two-dimensional ISI channels, *Proc. IEEE Int. Symp. Information Theory*, Adelaide, pp. 1348–1352, 2005.

8. E. Kurtas, J.G. Proakis, and M. Salehi, Coding for multitrack magnetic recording systems, *IEEE Trans. Information Theory*, **43**, pp. 2020–2023, 1997.

9. P.J. Davey et al., Two-dimensional coding for a multi-track recording system to combat intertrack interference, *IEEE Trans. Magn.*, **34**, pp. 1949–1951, 1998.

10. M. Kobayashi, J. Boutros, and G. Caire, Successive interference cancellation with SISO decoding and EM channel estimation, *IEEE J. Sel. Areas Commun.*, **19**, pp. 1450–1460, 2001.

11. W. Kozek and A. Molisch, Nonorthogonal pulse shapes for multicarrier communications in doubly dispersive channels, *IEEE J. Sel. Areas Commun.*, **16**, pp. 1579–1589, 1998.

*References marked with an asterisk are recommended as supplementary reading.

12. D. Dasalukunte, Multicarrier faster-than-Nyquist signaling transceivers, Ph.D. thesis, Electrical and Information Technology Dept., Lund University, Lund, Sweden, 2012; substantially reproduced as D. Dasalukunte et al., *Faster-than-Nyquist Signaling Transceivers: Algorithms to Silicon*, Springer, New York, 2014.

13. D. Dasalukunte, F. Rusek, and V. Öwall, Multicarrier faster-than-Nyquist signaling transceivers: hardware architecture and performance analysis, *IEEE Trans. Circuits Syst. I*, **58**, pp. 827–838, 2011.

14. B. Le Floch, M. Alard, and C. Berrou, Coded orthogonal frequency division multiplex, *Proc. IEEE*, **83**, pp. 982–996, 1995.

15. A. Alard, Construction of a multicarrier signal, U.S. Patent 6,278,686, 2001.

16. P.W. Dent, Method and apparatus for communication with root-Nyquist self-transform pulse shapes, U.S. Patent application 2009/0003472 A1, 2009.

17. I. Kanaras, A. Chorti et al., Spectrally efficient OFDM signals: bandwidth gain at the expense of receiver complexity, *Proc. IEEE Int. Conf. Commun.*, Dresden, 2009.

18. S. Isam and I. Darwazeh, Design and performance assessment of fixed complexity spectrally efficient FDM receivers, *Proc. IEEE Vehicular Tech. Conf.*, (Spring), Yokohama, 2011.

19. R.G. Clegg, S. Isam et al., A practical system for improved efficiency in frequency division multiplexed wireless networks, *IET Commun.*, **6**, pp. 449–457, 2012.

20. T. Xu and I. Darwazeh, Spectrally efficient FDM: Spectrum saving technique for 5G, *Proc. Int. Conf. on 5G for Ubiquitous Connectivity*, Akaslompolo, Finland, pp. 273–278, 2014.

21. P.N. Whatmough, M.R. Perrett et al., VLSI architectures for a reconfigurable spectrally efficient FDM baseband transmitter, *Proc. IEEE Int. Symp. Circuits and Systems*, Rio de Janeiro, pp. 1688–1691, 2011.

22. A. Barbieri, D. Fertonani, and G. Colavolpe, Time-frequency packing for linear modulations: spectral efficiency and practical detection schemes, *IEEE Trans. Commun.*, **57**, pp. 2951–2959, 2009.

23. G. Fettweis, M. Krondorf, and S. Bittner, GFDM – Generalized frequency division multiplexing, *Proc. IEEE Vehicular Tech. Conf. (Spring)*, Barcelona, 4 p., 2009.

24. W. Yuan et al., A graphical model based frequency domain equalization for FTN signaling in doubly selective channels, in submission, *IEEE Int. Symp. Personal, Indoor and Mobil Radio Communs.*, Valencia, 2016.

25. N. Pham, J-M. Freixe, and A. Bonnaud, EUROSAT, Paris, personal communication, 2013.

26. N. Pham, J-M. Freixe, A. Bonnaud et al., Exploring faster-than-Nyquist for satellite direct broadcasting, *Proc. Int. Conf. Satellite Commun.*, Florence, 2013.

27. M. Zhu, Q. Guo, B. Bai, and X. Ma, Reliability-based joint detection-decoding for nonbinary LDPC-coded modulation systems, *IEEE Trans. Commun.*, **64**, pp. 2–14, 2016.

6

CODED MODULATION PERFORMANCE

INTRODUCTION

It is important to compare FTN-style schemes to competing methods. Chapter 6 reviews several older coded modulations that achieve moderate bit densities in the range 2–4 bits/Hz-s. In the time of their development—the period 1975–1990—they were the most bandwidth efficient coding methods available, and the insight that coding could reduce energy without bandwidth expansion arose with them. Although these methods will not achieve the bit densities seen in the earlier chapters, they reach good energy efficiency for their densities and their receivers are usually simpler. They are thus still valuable.

A coded modulation is a scheme that encodes actual signal amplitudes and phases, rather than abstract symbol streams. Section 6.1 explores set-partition codes. These encode sequences of QAM carrier modulation symbols (QAM is introduced in Chapter 2). The set partition idea means that the QAM constellation is first broken into subsets. An encoder chooses which one; some databits select the subset and some are sent in the chosen subset without further coding. This innovation is important because it simplifies the coding and it gives a means to achieve higher bit densities. The method is often called TCM (trellis-coded modulation), but this term embraces older methods as well, including simple concatenation of a convolutional or other trellis code with QAM modulation. This direct approach has greater potential in theory but has seen less success in the narrowband scenario.

Bandwidth Efficient Coding, First Edition. John B. Anderson.
© 2017 by The Institute of Electrical and Electronics Engineers, Inc. Published 2017 by John Wiley & Sons, Inc.

Section 6.2 explores CPM (continuous phase modulation) codes. Also based on carrier modulation, CPM is an older method that adds a far reaching constraint: the transmitted signal has a constant envelope. The data is therefore sent only in the phase of the signals, and CPM adds the constraint that the phase must be continuous. Originally this was to reduce spectral sidelobes, but by adding further continuity constraints CPM can be attractive at quite narrow bandwidths however measured. Like FTN, CPM can be coded or uncoded. A more natural bandwidth measure for CPM is 99% or higher power-in-band (PIB), and the section introduces how to convert signal bit densities from one bandwidth criterion to another. The modulator pulses in FTN and set-partition coding have almost no spectral sidelobes, and the half power criterion is more useful for them, but separate design of sidelobes is an important part of CPM signaling.

6.1 SET-PARTITION CODING

The set partition concept divides a constellation of signal points into disjoint subsets. At each symbol interval, a subset is selected by a scheme such as a convolutional encoder. Some databits drive the selection and others are sent via the choice of a point in the subset. The idea of sequencing through subsets under the direction of a convolutional encoder was formally published by Ungerboeck [7] in 1982, although the paper was delayed and the idea appeared in earlier papers with attribution to Ungerboeck. Much research occurred in the 1980s and is summarized in the survey paper [8], and the textbooks by Proakis [3], Biglieri et al. [4], Schlegel and Perez [5,6], and in Reference 2, among other places.

This code-driven partition concept found extensive practical use, notably in telephone-line modems. The idea applies to lattices, PAM, and phase-shift key-ing (PSK) constellations, but almost all implementations work with straightforward QAM or PSK constellations. We de-emphasize the PSK schemes because they do not lend themselves to bandwidth efficient coding.

6.1.1 Set-Partition Basics

QAM itself is simple linear modulation. It requires the in-phase and quadrature signal description first given in Chapter 5, but without time or frequency FTN. This is

$$\sqrt{2E_s}\left[I(t)\cos 2\pi f_c - Q(t)\sin 2\pi f_c\right],$$

where

$$I(t) = C_0 \sum_{n=0}^{N-1} u_n^I h(t - nT)$$

and

$$Q(t) = C_0 \sum_{n=0}^{N-1} u_n^Q h(t - nT),\tag{6.1}$$

in which $h(t)$ is a unit-energy T-orthogonal pulse, T is the symbol time, the 2-tuples (u_n^I, u_n^Q) are M-ary valued, and C_0 is a normalizing constant that satisfies

$$C_0 = \left[(1/M) \sum_{u^I, u^Q} [(u^I)^2 + (u^Q)^2] \right]^{-1/2}.\tag{6.2}$$

Because of C_0, the expected value of $\int [I^2(t) + Q^2(t)]\, dt$ is always 1 and E_s is the QAM symbol average energy. Each QAM symbol is a point in a two-dimensional constellation formed from independent I and Q dimensions. For the only time in the book, M counts the values of a 2-tuple. The set of all 2-tuples that can appear in the codewords is called the *master constellation*. By convention, u^I and u^Q take values in the usual PAM alphabet $\{\pm 1, \pm 3, \dots, \pm \mu\}$, but all μ^2 possibilities may not be present in the master constellation. If they are, the constellation is *rectangular QAM*, and the MQAM transmission can then be viewed as two independent μPAM transmissions of the type in the earlier chapters, where $\mu = \sqrt{M}$.

The normalized Euclidean square distance between two MQAM signals $s_1(t)$ and $s_2(t)$ remains the form in Eq. (2.8), $(1/2E_b) \int [s_1(t) - s_2(t)]^2\, dt$, which for large carrier frequency f_c may be written as

$$\frac{\log_2 M}{2} \int [(\Delta I(t))^2 + (\Delta Q(t))^2]\, dt,\tag{6.3}$$

which may in turn be written

$$\frac{\log_2 M}{2} C_0^2 \sum_n [(\Delta u_n^I)^2 + (\Delta u_n^Q)^2].\tag{6.4}$$

As in previous chapters, the notation ΔX means the difference between two X.

Design of a set-partition code comprises three parts: Choosing the *master constellation*, forming the *subsets*, and designing the *subset selector* mechanism. Some databits drive the subset selection and the rest are carried in the chosen subset. Almost all QAM-based practical codes are based on a rectangular QAM master constellation, or a subset of one, which contains $M_c = 2^\kappa$ points, κ an integer. Rounding and compacting the QAM constellation, replacing it with a lattice, or extending the constellation over more than two dimensions can reduce E_b/N_0 up to 1 dB, but complexity rises steeply. These measures are seldom employed in practice and will not be discussed.

An effective method to form the subsets was given by Ungerboeck in Reference 7 and is illustrated in Figure 6.1 for 16QAM. Beginning with the master constellation,

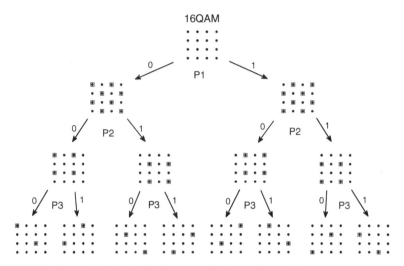

FIGURE 6.1 Set partitioning by the Ungerboeck rules, using 16QAM as an example. Squares indicate subset points. Each row has half the subset points. The "0" and "1" symbols are the subset selector outputs.

each subset is split to form two; effectively, each two children have alternate points from their mother. It is clear from the figure that J rows of splits produces 2^J subsets, each with $M_c/2^J$ points. Some examination will show that the square minimum distance in the subsets doubles each row, to become 2^J times the distance in the master constellation. The first partition does not lead to a useful result, so we have that $1 < J < \log_2 M_c$.

For a code that carries R bits per two dimensions, $R - b$ bits are carried in the subset and the remaining b bits select which subset it is. Many methods can be used, but virtually all set-partition codes utilize a rate b/c binary convolutional encoder. The parameter c must satisfy $R - b + c = \log_2 M_c$. Figure 6.2 shows the six configurations that are available when $M_c = 32$.[1] A similar chart with $\log_2 M_c - 2$ levels exists for each M_c that is a power of 2. There is one column for each databit rate R. Down each column the selector rate grows, the convolutional code becomes more complex, and there are more subsets with fewer points and higher minimum distance. The rightmost column, with $R = \log_2 M_c - 1$ bits per two dimensions, generally leads to the most satisfactory codes. Furthermore, for any M_c, the convolutional codes there are the familiar ones with rates $1/2, 2/3, 3/4, \ldots$.

The $d_{\min}$ of a set partition code is the minimum of two distances, the minimum distance of a subset, called d_{ss}, and the intersubset minimum distance, the worst case

---

[1] The standard 32-point constellation, called 32 Cross, is formed from rectangular 36QAM by removing one point from each corner. Similarly, 128 Cross is formed from 144QAM by removing 4 points from each corner. Since 32 and 128 lack integer square roots, these cannot be realized by the product of two PAMs.

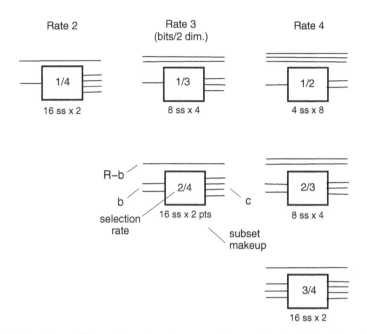

FIGURE 6.2 Six possible configurations for set-partition coding that is based on a 32-point master constellation. Parameter notation is as shown in the center.

between two sequences that start with different subsets. Each partitioning of subsets doubles d_{ss}^2; for example, d_{ss}^2 down the right-hand column of Figure 6.2 is 1.6, 3.2, and 6.4. With a sufficiently complex subset selector, the intersubset square distance can always be driven larger than d_{ss}, so that the game of code design is to set a selector complexity and split subsets until d_{ss} exceeds the selector capability. Good selectors for one M_c tend to be good for a range of M_c.

A final complication is that the mapping from the $\log_2 M_c$ bits to QAM points needs to be optimized. The search for a good code thus comprises finding a good master constellation, subset configuration, selector, and mapping. Outcomes of this complex search are discussed in Reference 2 and references therein.

In the definitions of Chapter 1, the mapping from R bits to the master constellation is the M_c-ary modulator, and a set-partition code is a true code since it works to select some but not all of the possible modulator outputs. It is clear that the performance potential grows as the subset size reduces. Conversely, it is also clear that the set partition idea *limits* performance, to that obtainable from d_{ss}. To remove the limit, one must drop the idea of subsets and feed all databits to an "encoder" block. This is simply a convolutional or other encoded QAM with a mapper. Nonetheless, set partitioning has solid advantages. It transfers some of the coding effort to the fixed subset arrangement, rather than placing it all, for example, on complicated convolutional coding. It also provides an avenue to reach very high bit density: Keep the convolutional code fixed and simply increase all the point sets.

Virtually all receivers for set-partition codes employ an adaptation of the Viterbi algorithm.

Spectrum and Bit Density. As with FTN coding, the baseband spectrum of a set-partition code is simply the power spectrum of the modulation pulse $h(t)$, subject to the usual assumption that the coding does not correlate the QAM symbols. Unlike FTN and the CPM in the next section, the only way to reduce spectrum is to increase the modulator alphabet. QAM is a passband modulation, so that the positive-Hz occupied spectrum corresponds to the left and right sides of the baseband spectrum, and this fact divides the databits per Hz-s by 2 (alternately, the I and Q dimensions must be accounted for). Using the half-power criterion, the databit density becomes

$$\frac{R}{2T B_3} = \frac{1}{2T_b B_3} \qquad \text{bits/Hz-s,} \qquad (6.5)$$

where R databits are carried per two-dimensional QAM symbol, T is the symbol time, T_b is the bit time, and B_3 is the baseband positive 3 dB down bandwidth of $h(t)$. As always, the last is $1/2T$ Hz for orthogonal pulses, which further simplifies the bit density to $R/(2T/2T) = R$ bits/Hz-s.

The $h(t)$ in applications have rapid spectral rolloff and so we continue to use half-power bandwidth as a bandwidth criterion. For the 30% root RC pulse, the 99% power in band frequency is $0.567/T$, which is 13% larger than B_3. With the 99% criterion, therefore, bit densities are 88% of their values with the half-power criterion.

6.1.2 Shannon Limit and Coding Performance

The ultimate Shannon limit for set-partition codes—or for methods without partitioning—stems from the PSD capacity C_{psd} found in Section 3.2. This limit appears on the performance plots to come. Several more constrained Shannon limits have received study as well, and are interesting because they give insight into the cost of restrictions on the coding design.

The traditional square-spectrum capacity C_{sq} leads to the Shannon limit for schemes based on orthogonal modulation pulses, which is the case in set-partition coding. This limit also appears on the performance plots. It can make schemes seem several dB closer to Shannon performance than they really are. Another approach to capacity calculation is illustrated in Figure 3.1: Capacity can be computed for codes constrained to the QAM master constellation points, with or without a requirement for uniform probabilities on the points. The figure shows the values applicable to 16QAM (i.e., two 4PAMs) and 64QAM (two 8PAMs) with no uniform probability requirement, with comparison to the square-spectrum capacity. There is some loss of capacity evident, and in particular, an M_c-point QAM constellation cannot have capacity larger than $\log_2 M_c$ bits per two dimensions.

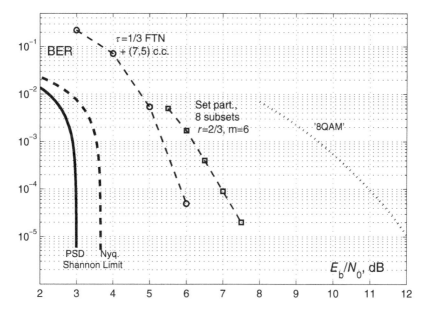

FIGURE 6.3   Rate 3 bits/Hz-s set-partition coding, compared to binary FTN with $\tau = 1/3$. The set-partition scheme has rate 2/3 memory 6 subset selection, 8 2-point subsets, 16QAM master constellation. Actual decoder test data adapted from Zhang [9]. Shannon limits at left; optimal 8-point QAM simple modulation performance at right.

To summarize, the capacity to which set-partition codes approach is reduced by pulse orthogonality, the QAM modulation, and the assumption of uniform points. The effect is to pull to the right the Shannon limit to the BER versus E_b/N_0 performance of the coding. The most significant source of loss is pulse orthogonality.

Error Performances. What follows is the BER performance of two set-partition codes, one at the moderate rate 3 databits/Hz-s and one at the high rate 6. Both work at rate $\log_2 M_c - 1$ databits per Hz-s and use complicated rate 2/3 recursive systematic convolutional subset selectors with many states. These features are chosen so that the codes represent the approximate best performance available. Substituting rate 1/2 selectors for the rate 2/3 moves error performance about 3 dB to the right, 3 dB further from the Shannon limit.

Figure 6.3 shows actual test results for a 16QAM master constellation, 8 subsets of 2 points each, and the 64-state rate 2/3 recursive systematic selector $(h_0, h_1, h_2) = (101, 016, 064).^2$ This configuration represents the most complex and highest performing set-partition code at 3 bits/Hz-s. On the left are the PSD and orthogonal-pulse ("Nyq.") Shannon limits, and on the right is the of the optimal

---

[2] The right-justified observer-form notation here for a rate 2/3 recursive systematic encoder is standard in the field. See Reference 2, p. 153, [5], Chapter 3, or Reference 6, Chapter 3.

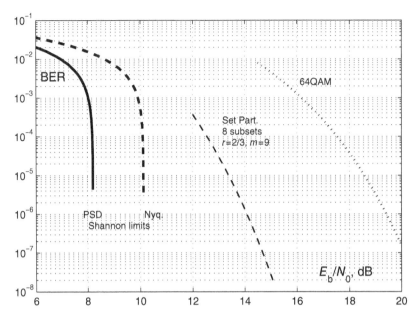

FIGURE 6.4 Rate 6 bits/Hz-s set-partition coding. Rate 2/3 memory 9 subset selection, 8 16-point subsets, 128 Cross master constellation. Estimated BER based on d_{min} and error event studies. Shannon limits at left; rectangular 64QAM simple modulation at right. Compare Figure 4.16.

8-point simple QAM modulator.[3] The set-partition scheme lies ≈ 4.7 dB from the PSD Shannon limit at BER 10^{-5} and improves on the simple modulation benchmark by ≈ 4.3 dB. Shown for comparison is a test of binary time-FTN with $\tau = 1/3$ and the (7,5) rate 1/2 convolutional encoder, a scheme with the same bits/Hz-s. It performs about 1.5 dB better. Its 4-state encoder is much simpler than the subset selector, but the FTN requires iterative decoding. We can assume that the complexity of the two schemes is roughly the same.

Figure 6.4 shows a very high rate, 6 bits/Hz-s. The master constellation is 128 Cross (see footnote after Eq. (6.4)), there are 8 subsets of size 16, and the selector is the rate 2/3 recursive systematic encoder $(\boldsymbol{h}_0, \boldsymbol{h}_1, \boldsymbol{h}_2) = (1001, 0346, 0510)$. Actual test results for such a complex scheme and high rate are difficult to come by, and d_{min}^2 (which is 1.17) produces a very weak BER estimate, caused by a multitude of long trellis decoding error events that lie at or near d_{min}. Nonetheless, it is important to attempt a comparison to the main competitor, 4-ary FTN. Schlegel and Zhang (see References 5 and 9) have made a study of such error events, and based on their findings, we can estimate the BER to be $\approx 45Q(\sqrt{1.17E_b/N_0})$. This produces an outcome about

---

[3]No rectangular QAM has 8 points. The benchmark here has points at $\{(\pm 1, \pm 1), (0, \pm(1 + \sqrt{3})), (\pm(1 + \sqrt{3}), 0)\}$, symbol error rate $\approx 3.5Q(\sqrt{1.27E_b/N_0})$, and BER a little higher.

5 dB from the PSD Shannon limit and 5 dB better than the 64QAM simple modulation benchmark, not too different from the rate 3 bits/Hz-s tests. Three FTN schemes at this rate are shown in Figure 4.16. The middle of these reaches BER 10^{-8} at $E_b/N_0 = 12$ dB, about 3 dB closer to the Shannon limit than the set-partition code.

Observe that fully 2 dB is lost at 6 bits/Hz-s by basing the Shannon limit on orthogonal pulses. Researchers continue to search for better combinations of subset, selectors, and mappings; see for example Reference 6, Chapter 3.

The ultimate problem, trellis-coded QAM and PSK without subsets, has attracted attention since the 1960s. Codes are known at only relatively low rates. Many rate 3 16QAM and rate 2 8PSK codes are due to Zhang [6,9] and to Porath [14].

6.2 CONTINUOUS PHASE MODULATION

CPM arose in the 1970s and was the first coded modulation to be extensively researched. It was here that the idea of narrowband coding and practical coding gains without bandwidth expansion first arose. Single-chapter expositions about CPM appear in References 2, 3, and 1 is a book-length treatise. The early history of CPM is recounted in Reference 2, Chapter 5.

Like set-partition schemes, CPM schemes are bandpass, with a carrier frequency. They differ from the other methods in the book in that they are strictly constant-envelope: They encode data in the *phase* of a signal. They also employ nonlinear modulation. This adds complexity to energy and bandwidth calculations. Unlike linear modulation, there is no pulse that sets the spectrum; spectrum and energy efficiencies interact and only general rules can be given. The standard CPM transmitter convolves data with a generator to obtain its phase much like a convolutional encoder, and CPM is often considered a type of coding, but we will consider it as a modulator. The signals so modulated can be coded or not; some would consider coded CPM to be a form of concatenated coding.

The first application of CPM was satellite communication but focus soon shifted to mobile radio. Early digital satellites and the 2nd Generation GSM cellular telephone system both featured simple CPM systems. In these applications, the aim was to reduce spectral sidelobes and thus reduce cochannel interference, while not sacrificing energy efficiency.

6.2.1 CPM Basics

As in set-partition coding, CPM signals take the I and Q form in Eq. (6.1), but since data is carried in the phase only, a more convenient form is

$$s(t) = \sqrt{2E_s/T} \cos[2\pi f_c t + \phi(t, \boldsymbol{u})], \qquad (6.6)$$

in which the *excess phase* $\phi(t, \boldsymbol{u})$ carries the databits in the sequence of M-ary modulation symbols $\boldsymbol{u}$. Here T is the symbol time and E_s is the energy devoted to each u for large f_c, not the energy in a 2-tuple as in Section 6.1; this can be verified by

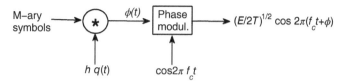

FIGURE 6.5   Standard modulator for CPM signaling.

finding $\int_0^T s^2(t)\,dt$. The signal $s(t)$ has I and Q components and the distance between two signals satisfies Eq. (6.3), but a more convenient distance formula in terms of phase only is

$$d^2 = \frac{\log_2 M}{T} \int_{\mathcal{I}} [1 - \cos \phi(t, \Delta \boldsymbol{u})]\, dt, \qquad (6.7)$$

where $\mathcal{I}$ is the interval over which the phase signals differ, and Δ denotes the difference of two streams. The constant-envelope condition means that $I^2(t) + Q^2(t)$ is a constant for all t.

A block diagram of CPM is shown in Figure 6.5. A CPM method is specified by its excess phase function $\phi(t, \boldsymbol{u})$. This is in turn specified by a phase response $q(t)$, modulation index h, and modulation alphabet of size M. In standard CPM, these are subject to the following conditions:

(i) The excess phase satisfies the convolution

$$\phi(t, \boldsymbol{u}) = 2\pi h \sum_n u[n] q(t - nT) \qquad (6.8)$$

in which

(ii) The phase response is continuous and satisfies $q(t) = 0, t < 0$ and $q(t) = 1/2$, $t > LT$, where $L \geq 1$ is an integer and T is the M-ary symbol time; and

(iii) The symbols $u[n]$ take values in the set $\{\pm 1, \pm 3, \ldots, \pm M\}$.

The function $q(t)$ is the generator in a convolution. The quantity $2\pi h$ is the total eventual phase change caused by a shift of u to a neighboring value. When $L = 1$, the CPM is called full response, and all such phase change occurs in one symbol interval; when $L > 1$, the CPM is partial response.

The derivative of $q(t)$ is the instantaneous frequency offset to the carrier. The conditions on $q(t)$ guarantee phase continuity, an essential requirement for a good spectrum. Some important phase responses are shown in Figure 6.6. Example (a) has the common name CPFSK (continuous-phase frequency-shift keying), since the linear phase corresponds to a constant frequency shift $hu[n]/2$ Hz from the carrier. Figure 6.6b is the same $q(t)$ but partial response, covering $L = 2$ intervals; Figure 6.6c is a full-response $q(t)$ called 1RC, meaning a raised-cosine response shape on one interval; Figure 6.6d is called 3RC, the same shape but spread over three intervals.

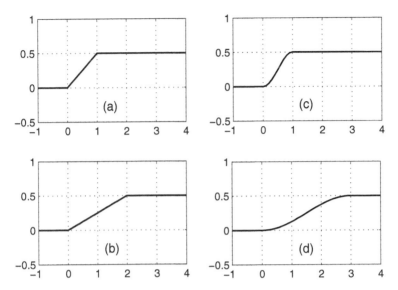

FIGURE 6.6 Four examples of the phase response function $q(t)$: (a) full response CPFSK; (b) partial response CPFSK, two intervals; (c) full response RC; (d) partial response RC, three intervals.

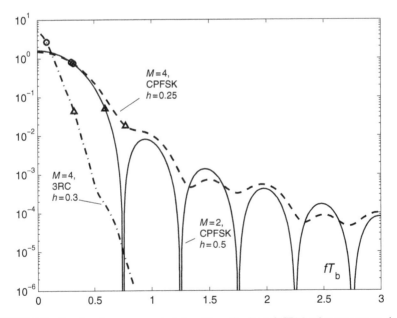

FIGURE 6.7 Baseband power spectral densities of selected CPM schemes versus databit normalized frequency. (a) CPFSK with $M = 2, h = 0.5$; (b) CPFSK with $M = 4, h = 0.25$; (c) 3RC with $M = 4, h = 0.3$. Triangles are 99% PIB bandwidths; circles are 3 dB bandwidths.

Achieving a narrow PSD breaks into two problems: a narrow main lobe and rapidly declining sidelobes. The first is achieved primarily by a longer partial response width L, and secondarily by a larger modulation size M. The rate of decline of the sidelobes is set by Baker's Rule, which states that the PSD rolls off as $|(f - f_c)T|^{-2\zeta-4}$, where ζ is the number of continuous derivatives in $q(t)$ (see Reference 2, p. 65). The smoothness of $q(t)$ is thus critical to spectral measures like the 99% PIB. CPFSK with any L or M obeys $|(f - f_c)T|^{-4}$, which is 12 dB/octave. The RC phase pulse schemes obey $|(f - f_c)T|^{-8}$. The difference is obvious in Figure 6.7, which compares the PSDs of three representative CPM schemes, namely binary CPFSK,[4] 4-ary CPFSK, and 4-ary 3RC: The first two have similar sidelobe spectra, while 3RC rolls off much faster. The spectra here[5] are normalized to the databit time T_b, and 99% PIB values are shown by triangles and half-power bandwidths by circles. The long partial response in the 3RC system leads to a very narrow main lobe.

As with any signal set, the receiver error rate of a CPM scheme is given approximately by $Q(\sqrt{d_{\min}^2 E_b/N_0})$, where $d_{\min}$ is the minimum distance of the signal set and E_b is the average energy per databit. Distance does not have a simple relation to q, L, h, M, although it can be said that distance generally grows with the response length L and shrinks with the index h. From a combined energy–bandwidth point of view, the best modulation M lies in the range 4–8. For CPFSK with reasonable h ($h \log_2 M < 1$), a simple relation holds:

$$d_{\min}^2 = 2 \log_2 M \left[1 - \frac{\sin 2\pi h}{2\pi h} \right]. \tag{6.9}$$

Otherwise, the calculation is more complex (see References 1 and 2) and $d_{\min}^2$ can lie considerably above and below the benchmark QPSK value of 2, depending on the bandwidth. Because of nonlinearities, distance can depend on past symbol history. But this can be a benefit since $d_{\min}$ applies only after certain histories, and consequently the receiver BER generally lies in the range 0.5–2 times the Q-function.

CPM is a trellis modulation and is most often decoded by the Viterbi algorithm. Reduced-search algorithms (Section 4.2.2) work well; see Reference 2, Section 5.3.3. Within the framework of the book, CPM is a modulation, since it produces signals from all possible M-ary inputs. We will look at coded CPM, which transmits some but not all of these in an encoded pattern. Nonetheless, uncoded CPM includes the phase convolution, and to many it thus seems coded.

---

[4]A popular system in constant-envelope applications, binary CPFSK with $h = 0.5$ is also known as MSK (minimum-shift keying) or fast FSK. It is relatively wide band, but much narrower than square pulse QPSK, which has 99% PIB some five times larger. Formulas for the CPFSK PSD appear in Reference 1, p. 167, Reference 2, p. 64, and Proakis [3].

[5]In this section, the PSD is taken as the baseband average power spectrum of a bandpass signal, see Reference 3. The full bandpass PSD at positive Hz can be taken as the positive and negative sides of the baseband spectrum.

6.2.2 Bits per Hz-s and the Shannon Limit in CPM

As with set-partition codes, CPM is a double-dimension passband signal. Its databit density satisfies Eq. (6.5), except that its (uncoded) rate R is now $\log_2 M$, where M is the size of the phase modulation in Eq. (6.6). The density becomes

$$\frac{\log_2 M}{2T B_{\text{crit}}} = \frac{1}{2T_b B_{\text{crit}}} \qquad \text{bits/Hz-s,} \qquad (6.10)$$

where T is the CPM M-ary symbol time and B_{crit} is with respect to a bandwidth criterion such as half power or 99% PIB. Since M is single dimensional, it is the square root of the QAM point count, and the density expression has a handicap of more or less a factor $\log_2 M$ relative to one based on QAM. However, the spectrum shape is part of the CPM design process, and the PSD can be designed to be more narrow than that of QAM.

A subtlety in this discussion is the fact that CPM signals are constant envelope. Class C and similar power amplifiers for constant-envelope carrier signals are 2–4 dB more efficient than the Class A or B that are needed for set-partition and FTN signals. When there is a limited battery supply, as with satellites and handheld terminals, power amplifier efficiency dominates and this 2–4 dB should really be subtracted from the E_b/N_0 required for CPM. This approximately cancels the single-dimensionality loss, or alternately, the loss due to the constant envelope requirement.

Unlike FTN and set-partition coding, the 99% PIB frequencies for CPM are much larger than the half-power values. Instead of the 13% extra that applies to 30% root RC pulses, Figure 6.7 shows 99% PIBs that are two to four times larger. It needs to be emphasized that no matter what bandwidth criterion is chosen, the Shannon limit is the same, because it depends on the entire spectrum. What changes is the rate parameter applied to the limit calculation, since this is in bits per Hz and second, and the Hz value depends on the criterion. The point is subtle, and we give an example of how to convert from one criterion to another.

Example 6.1 *(Bandwidth Criterion Conversion)*
From Figure 6.7, 4-ary CPFSK with $h = 0.25$ has half power (3 dB down) and 99% baseband bandwidths $B_3 T_b = 0.315$ and $B_{99} T_b = 0.775$ Hz-s, normalized to the databit time. The signals are considered to carry $\log_2 M = 2$ bits per symbol in twice these positive bandwidths. Take first the 3 dB criterion. Program gber-cap(rate,h,f) in Appendix 3A computes the Shannon limit for signals having rate bits/Hz-s and PSD h on positive frequencies f. However, the program assumes that the rate parameter and the PSD are with reference to the applicable B_{crit} set to 1 Hz. This is enforced by scaling up both by a factor $1/0.315$; the rate becomes $\log_2 M/2T B_3 = 2/2(0.315) = 3.174$ bits/Hz-s and f is expanded by $1/0.315$. gbercap(rate,h,f) finds that $E_b/N_0 = 1.5$ dB for BER = 0, and gives the BER versus E_b/N_0 Shannon limit shown in Figure 6.8a. The total P/N_0 required when B_3 is scaled up to 1 Hz is 4.52. Note that both rate and total power are expanded by the same factor, so that E_b is unchanged. If instead we take the 99% criterion, the

scale-up factor is $1/0.775$ and `rate` becomes $2/2(0.775) = 1.290$ bits/Hz-s. This time `gbercap(rate,h,f)` finds again that $E_b/N_0 = 1.5$ dB for BER = 0 and computes the same Shannon limit. This will be true whenever `rate` and `f` are scaled by the same factor.

CPM Capacity. The ultimate Shannon capacity against which CPM signals can be compared is C_{psd} in Chapter 3, the capacity computed for their PSD, and this capacity is used in the Shannon limit program `gbercap(rate,h,f)`. This comparison is perhaps too strenuous, since C_{psd} does not take into account constraints such as constant envelope, the phase pulse shape $q(t)$, and the alphabet M, among others. A full calculation is challenging because the signals are Markov chains and the effective channel is not memoryless.

Nevertheless, a related quantity, the *computational cutoff rate* R_0 can be computed using a special eigenvalue technique (first applied in Reference 10 and summarized in Reference 1, Section 5.6, and Appendix B). The utility of R_0 here is that it is a rather good lower bound to capacity. The details of the method are beyond our scope but the outcomes can be summarized. The cutoff rate, and very likely capacity, depends on the modulation index h, with different h optimal in different ranges of the $(E_b/N_0, 2BT_b)$ plane.[6] Longer phase pulses and larger M have higher cutoff rates than smaller ones.

6.2.3   Error Performance of CPM

Figure 6.8 shows the BER of two uncoded 4-ary CPM schemes, along with their respective Shannon limits, plotted on the same scales. Both BERs are taken as $\approx 1.5\,Q(\sqrt{d_{min}^2\,E_b/N_0})$; the small leading coefficient is typical of actual decoder tests. The phase pulses and spectra of the CPMs appear in Figures 6.6 and 6.7.

Figure 6.8a shows CPFSK plot, $h = 0.25$, which has $d_{min}^2 = 1.45$, $B_3 T_b = 0.315$ Hz-s, and $B_{99}T_b = 0.775$. Its error performance lies quite far from the Shannon limit for its bit density, which is 3.17 bits/Hz-s relative to B_3 and 1.29 relative to B_{99}. The position may be due to the simplicity of CPFSK, which acts like a simple modulator.

Figure 6.8b is a much more complex CPM, a 3RC phase pulse with $h = 0.3$, $d_{min}^2 = 1.4$, $B_3 T_b = 0.082$, and $B_{99}T_b = 0.322$. Its error performance lies much closer to the Shannon limit for its bit density, which is 12.2 bits/Hz-s relative to B_3 and 3.11 relative to B_{99}. These densities and the rightward position of the limit reflect the very narrowband nature of the signaling. This CPM has a phase response that is both three times longer and much smoother than the first one. Its performance and Shannon limit do not differ much from the density 6 set-partition code in Figure 6.4.

Coded CPM. CPM was exhaustively studied in the period 1975–1990, and some of this work investigated convolutional coding and single-interval CPFSK. This

---

[6]A version of this plane and an unconstrained capacity appear in Figure 3.2.

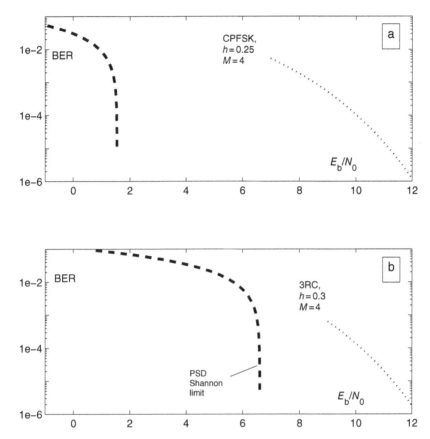

FIGURE 6.8 BER versus E_b/N_0 for two uncoded CPM schemes, compared to Shannon limits derived from their PSDs. (a) 4-ary CPFSK, $h = 0.25$, $d_{min}^2 = 1.45$; (b) 4-ary 3RC, $h = 0.3$, $d_{min}^2 = 1.4$. Axis scales are the same in both.

combination was meant to compete with early MSK-like systems and the outcomes are not striking from a modern combined energy–bandwidth point of view. But there were a few studies reported by Lindell et al. [11,12] which studied longer and much smoother phase response functions. When these are combined with a small modulation index, a large phase alphabet, and a high-rate convolutional code, the result can be an energy-efficient narrowband coding system with a bit density near 4 bits/Hz-s.

A BER estimate for one such system is shown in Figure 6.9, together with the Shannon limit that corresponds to its spectrum. The CPM modulation here uses a phase response $q(t)$ with a raised-cosine shape spread over two symbol intervals, combined with a small modulation index $h = 1/11$ and a large 8-ary modulation alphabet. The databits drive a feedforward semisystematic rate 2/3 convolutional encoder followed by an octal mapper that selects the modulation symbol. Both encoder and mapper were chosen after much searching, and together with the CPM

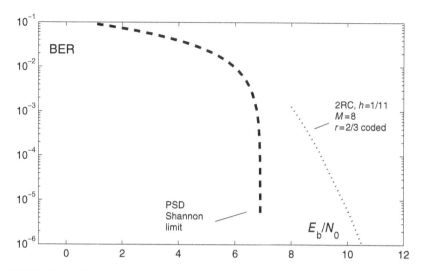

FIGURE 6.9 BER versus E_b/N_0 for a coded CPM scheme, compared to Shannon limit derived from its PSD. Convolutional encoder rate is 2/3, memory 4. CPM is $L = 2$ raised-cosine $q(t)$, $M = 8$, $h = 1/11$, with $B_{99}T_b = 0.275$ Hz-s. Scheme bit density is 3.6 bits/Hz-s relative to 99% PIB criterion.

modulator they create a transmitted signal set with square minimum distance 2.12 [12]. The 99% PIB normalized to the databit rate is $B_{99}T_b = 0.275$ Hz-s/bit, which is 3.6 bits/Hz-s. Using the left justified notation in Appendix 4B, the encoder matrix is $\begin{bmatrix} 40 \ 00 \ 00 \\ 00 \ 10 \ 66 \end{bmatrix}$. Very few receiver tests are available for non-CPFSK systems, and so we must estimate BER performance by means of $A\,Q(\sqrt{d^2_{\min}\,E_b/N_0})$.[7]

The result is Figure 6.9. It can be compared to the time-FTN tests in Figure 4.15. That figure shows 4 bits/Hz-s systems, relative to the half-power bandwidth criterion; for the 30% RC spectrum there, the 99% PIB criterion implies a 13% higher measure of bandwidth, which gives $\approx$ 3.5 bits/Hz-s, about the same density value as the CPM scheme. The FTN performance requires E_b/N_0 in the range 6–7.5 dB, 1–2 dB less than the CPM scheme. The 3 bits/Hz-s set-partition code in Figure 6.3 has roughly similar performance to the CPM when one accounts for the difference in bit density. These comparisons are the more attractive for coded CPM when one remembers that the CPM is constant-envelope.

It is interesting to compare this coded CPM system to Figure 6.8b, which has a roughly similar bit density (3.1 bits/Hz-s) but requires 1–2 dB more E_b/N_0, despite its somewhat lower bit density. We can view this 1–2 dB as coding gain.

---

[7]From the tests that are available [12,13], we can set $A \approx 10$ when the BER is 10^{-2}–10^{-4}, declining to $A \approx 1$ at 10^{-6}. The coding, small h, large M and $L = 2$ create quite a large A at middle BER.

6.3   CONCLUSIONS FOR CODED MODULATION; HIGHLIGHTS

The results collected in this chapter show that coded modulation can perform rather well in a combined energy–bandwidth sense, although it is most versatile in the 3–4 bits/Hz-s bit density range. Careful design is needed so that the parameter choices promote bandwidth efficiency: CPM requires high-rate coding, partial-phase response, a large alphabet, and a small modulation index; set-partition coding requires a complex, high-rate subset selector. CPM is special because it is constant envelope, and in some applications this is worth several dB in E_b/N_0, but in an application where a linear channel is a given, the constant envelope just wastes bandwidth efficiency. It appears that set-partition coding at high bit densities suffers several dB loss because it does not work toward the full nonorthogonal pulse capacity C_{psd}; CPM, on the other hand, does not necessarily lose this advantage. It would be interesting in future work to investigate set-partition codes based on nonorthogonal pulses.

It is difficult to make precise comparisons among coding methods that employ such different computations. FTN methods today are based mostly on iterative decoding. Set-partition coding avoids this complication, but competitive subset selectors are quite complex, and large-alphabet QAM is required, while FTN works well with a four-letter alphabet. CPM signals demand more careful synchronization.

More evaluation needs to be performed for some of the newer coding methods, including especially LDPC and BICM (bit-interleaved coded modulation). LDPC has been compared somewhat (see Section 5.3). Until now these have been low energy—wideband methods, but they may have interesting narrowband extensions.

REFERENCES

1. *J.B. Anderson, T. Aulin and C.-E. Sundberg, *Digital Phase Modulation*, Plenum, New York, 1986.

2. *J.B. Anderson and A. Svensson, *Coded Modulation Systems*, Kluwer-Plenum, New York, 2003.

3. *J.G. Proakis, *Digital Communication*, 4th and later eds., McGraw-Hill, New York, 1995.

4. *E. Biglieri, D. Divsalar, P.J. McLane and M. Simon, *Introduction to Trellis-Coded Modulation with Applications*, Macmillan, New York, 1991.

5. *C. Schlegel, *Trellis Coding*, IEEE Press, New York, 1997.

6. *C. Schlegel and L.C. Perez, *Trellis and Turbo Coding*, 2nd ed., Wiley, Hoboken, NJ, 2015.

7. G. Ungerboeck, Channel coding with multilevel/phase signals, *IEEE Trans. Inf. Theory*, **28**, pp. 55–67, 1982.

8. *G. Ungerboeck, Trellis-coded modulation with redundant signal sets, Parts I–II, *IEEE Communications Mag.*, **25**, pp. 5–21, 1987.

*References marked with an asterisk are recommended as supplementary reading.

9. W. Zhang, Finite-state machines in communications, Ph.D. thesis, Digital Communications Group, Univ. South Australia, Adelaide, 1995.

10. J.B. Anderson, et al., Power-bandwidth performance of smoother phase modulation codes, *IEEE Trans. Commun.*, **29**, pp. 187–195, 1981.

11. G. Lindell and C.-E. Sundberg, Multilevel continuous phase modulation with high rate convolutional codes, *Conf. Rec.*, IEEE Global Communs. Conf., pp. 30.2.1–30.2.6, 1983.

12. G. Lindell, On coded continuous phase modulation, Ph.D. thesis, Telecommunication Theory Dept., University of Lund, Sweden, 1985.

13. G. Lindell, personal communication, Feb. 2016. See also the following technical reports, listed in order of relevance and available from the library of Lund Univ., Sweden: T. Andersson, A limited tree search algorithm for noncoherent convolutionally encoded continuous phase modulation, Report TR-212, Teletransmission Theory Dept., Lund Tech. Univ., Oct. 1988; G. Lindell and C.-E. Sundberg, Error probability of coded CPM with Viterbi detection—Bounds and simulations, Report TR-190, *ibid.*, 1984; G. Lindell and C.-E. Sundberg, High rate convolutional codes and multilevel continuous phase modulation, Report TR-180, *ibid.*, 1983.

14. J.-E. Porath, On trellis coded modulation for Gaussian and bandlimited channels, Ph.D. thesis, Computer Engineering Dept., Chalmers Tech. Univ., Göteborg, Sweden, 1991.

7

OPTIMAL MODULATION PULSES

INTRODUCTION

Chapters 4 and 5 dealt with classical FTN, in which the linear modulator pulse $h(t)$ was an accelerated version of a T-orthogonal pulse with the new symbol time τT, $\tau < 1$. The favored pulse was a root RC with 10–30% excess bandwidth and the criterion for bandwidth measure was half power. While many applications would prefer a bandwidth criterion like 99 or 99.9% power in band, the RC spectrum does have a sharp cutoff and consequently these criteria are not much different.

Even so, this classical approach leaves at least three issues open.

- Are there significantly better base pulses than those that are T-orthogonal for some T, like the root RC pulse? While some pulses are not orthogonal, *all* linear modulation signal sets have the attractive Mazo limit property, even if no scaling of the base $h(t)$ satisfies orthogonality. The limit occurs because under increasing acceleration another error event eventually has distance less than the antipodal signal distance d_0, a value that upper bounds the minimum distance of any M-ary modulation signal set. An important example is the Gaussian pulse, which has attractive time and frequency properties but is not T-orthogonal for any T.

- Which base pulses and encoders are best for a power in band (PIB) criterion, such as 99%? Are they much different from pulses designed for a half-power criterion? Not only is the PIB criterion the important one in some applications,

Bandwidth Efficient Coding, First Edition. John B. Anderson.
© 2017 by The Institute of Electrical and Electronics Engineers, Inc. Published 2017 by John Wiley & Sons, Inc.

but also a sharp cutoff pulse like the root RC has unavoidable spectral side-lobes in a practical implementation, which can move a real-life PIB bandwidth considerably outward. Does this outward migration of energy have a significant effect?

- Every real symmetrical pulse $h(t)$ has a symmetrical Fourier transform $H(f)$. By Fourier transform duality, there is thus a *dual*-time pulse $H(t)$ that has transform $h(f)$. What is offered by these base pulses?

This chapter explores several ways to place these questions on a more secure footing. Section 7.1 treats *Slepian's Problem*: Given that all realizable pulses $h(t)$ must have a little energy outside of any given time and bandwidth interval, what pulse minimizes this energy? The answer is a prolate spheroidal wave pulse, and it has attractive modulation properties, in addition to its time–bandwidth optimality. An interesting outcome is that when a finite packet of data is transmitted, sinc and other orthogonal pulses with very small excess bandwidth are not optimal. Another outcome is that the Gaussian pulse is close to the prolate pulse; since it has good FTN properties as well, it is an important pulse to consider.

A second view of optimality, in Section 7.2, has been developed by Said: For an amount of power, say 1 or 0.1%, outside of a given bandwidth, which $h(t)$ maximizes the minimum distance? This distance d_{min} is important because it plays a primary role in how quickly the ISI due to $h(t)$ is reduced. Said's pulses are more practical to implement than root RC, and they have noticeably higher d_{min}. However, they do not necessarily lead to better narrowband coding, because other quantities than d_{min} affect performance.

The PSWF, Gauss and Said pulses will be applied to FTN transmission.

7.1 SLEPIAN'S PROBLEM

We think of physical signals as occupying finite time and bandwidth. Yet a signal or pulse with finite duration must have infinite spectrum width, and a finite spectrum must lead to infinite duration. Here lies a contradiction in our thinking. David Slepian thought long about this puzzle, and in a landmark 1976 paper [1] proposed the following resolution. Whether as humans or machines, we perceive signals only when they lie above a certain threshold, and their time and bandwidth should be measured relative to that. Let γ designate this fraction of the signal's energy. The signal time and bandwidth are the values outside which lie the fraction γ.

This is power-out-of-band (POB) thinking, first introduced in Section 1.3. It is further argued there that the more fundamental resource is the *product* of time and bandwidth in Hz-s, what we call the *signal occupancy* Υ. This follows because a signal's time and bandwidth can in principle be traded at will, by simply scaling time, and if time and bandwidth are measured by similar principles, Υ remains fixed.

For a modulation pulse $h(t)$, Slepian's problem becomes: For a fraction γ, which pulse minimizes occupancy? The answer is the prolate spheroidal wave function (PSWF). The body of mathematics supporting this forms a rich and subtle theory.

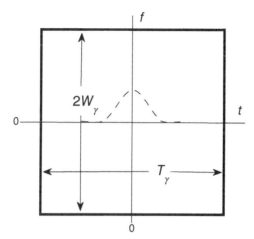

FIGURE 7.1   Definition of $\mathcal{T}_\gamma$ and $\mathcal{W}_\gamma$ that lead to γ out of band power for a pulse centered at time 0.

It developed in a series of papers by Slepian, Landau, and Pollak [2]–[7], of which the first two and the last are most germaine. We have space to mention only a few highlights and give some concrete results, the most important of which is the minimizing pulse. The PSWF was first applied to FTN in Reference 8.

7.1.1   PSWF Pulse Solution

The goal is to minimize the occupancy Υ. There are in reality two problems: Which pulse $h(t)$ has least Υ, and once pulses are combined in a linear modulation, possibly with FTN accelerations, what minimizes the Υ of the whole transmission? We take up the second problem in Section 7.1.3.

For the isolated pulse, we can assume that dual pulses have the same occupancy, since the frequency band for one is the time band for the other. Furthermore, scaling $h(t)$ in time scales $H(f)$ in frequency by the reciprocal factor, so that Υ and the fundamental shapes are unchanged. In much of Gaussian-noise communication and Shannon theory, only Υ matters, and scaled pulses are in a sense equivalent. In practical applications, of course, time and frequency can have rather different implications. One may wish to consume more of one than the other, or retain certain properties in one. Root RC pulses, for example, can be orthogonal in time, and they are confined in frequency but widely spread in time. If one wished the opposite, spread in frequency but confined in time, the dual to the root RC would be more attractive.

The task is then to derive the least-occupancy isolated $h(t)$ that is centered in the time–frequency box in Figure 7.1. The time and bandwidth of interest are $[-\mathcal{T}_\gamma/2, \mathcal{T}_\gamma/2]$ seconds and $[-\mathcal{W}_\gamma, \mathcal{W}_\gamma]$ Hz, the values that lead to the energy fraction γ out of band, including that above and below in frequency and left and right in time. Because $h(t)$ is symmetric, the energy is split $\gamma/2$ on the respective sides.

The plan is to find PSWFs, then find the one that corresponds to the out of band fraction γ, then derive the pulse from that PSWF. By convention PSWFs are defined in terms of the centered time duration $[-\mathcal{T}_\gamma/2, \mathcal{T}_\gamma/2]$ seconds and the positive bandwidth $[0, \mathcal{W}_\gamma]$ Hz, and the occupancy is taken as $\mathcal{W}\mathcal{T}$. For any $\mathcal{T}$ and $\mathcal{W}$, a PSWF is defined to be an *eigenfunction of the integral*

$$\lambda_i \psi_i(t) = \int_{-\mathcal{T}/2}^{\mathcal{T}/2} \frac{\sin 2\pi \mathcal{W}(t-s)}{\pi(t-s)} \psi_i(s)\, ds, \qquad \text{all } t, \ i = 0, 1, \ldots \qquad (7.1)$$

Here the set $\{\psi_i\}$ are eigenfunctions of the integral operator in the formula and $\{\lambda_i\}$ are their eigenvalues. We are interested only in $\psi_0(t)$, the principal eigenvector (the one with largest eigenvalue). It leads to the desired pulse shape and λ_0 to the in-band power. Important properties of $\psi_0(t)$ include the following. Proofs are given in Reference 2, p. 45ff, Reference 3, p. 65–80, and Reference 7.

- λ_0 depends only on the *product* $\mathcal{W}\mathcal{T}$. $\psi_0(t)$ depends only on the product in the sense that scaling $\mathcal{T}$ by a constant scales $\psi_0(t)$ in time by the same, and scales $\mathcal{W}$ and the Fourier transform $\Psi_0(f)$ in frequency by the inverse constant.

- As $\mathcal{W}\mathcal{T}$ grows, $\psi_0(t)$ tends to the Gauss pulse

$$g(t) = (1/\sqrt{2\pi a^2}) \exp(-t^2/2a^2), \qquad a^2 = \mathcal{T}/4\pi\mathcal{W} \qquad (7.2)$$

in the region $[-\mathcal{T}/2, \mathcal{T}/2]$. The Fourier transform is

$$G(f) = (1/\sqrt{2\pi\sigma^2}) \exp(-t^2/2\sigma^2), \qquad \sigma^2 = \mathcal{W}/\pi\mathcal{T}. \qquad (7.3)$$

- Because $\psi_0(t)$ is the principal eigenfunction, iteration of Eq. (7.1) in principle solves for $\psi_0(t)$; that is, an approximate $\psi(t)$ in the right-hand integral leads to a closer approximation on the left and so on, until a satisfactory convergence occurs.

Figure 7.2 plots the principle eigenfunctions when $\mathcal{T} = 1$ for the products $\mathcal{W}\mathcal{T} = 0.86, 1.28, 2.06$, which have eigenvalues near $0.9604, 0.9960$, and 0.99996. The Gaussian approximation Eq. (7.2) is shown for the $\mathcal{W}\mathcal{T} = 0.86$ case, and the approximation is reasonable in $[-1/2, 1/2]$. It is much more accurate for the other $\mathcal{W}\mathcal{T}$ (and invisible to see) because the product is larger. We will return to these solutions in Example 7.1.

It remains to derive the pulse shape and the out of band energy from $\psi_0(t)$ and λ_0. This requires a careful argument first given in Reference 3, pp. 65–80, which we can only summarize here. We take the special case where $1 - \gamma$ is the fractional energy inside *both* $[-\mathcal{T}/2, \mathcal{T}/2]$ and $[-\mathcal{W}, \mathcal{W}]$. Then the required eigenvalue is

$$\lambda_0 = \left[\cos \left(2 \cos^{-1}(\sqrt{1-\gamma}) \right) \right]^2. \qquad (7.4)$$

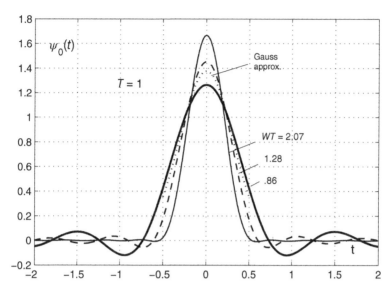

FIGURE 7.2   Eigenfunctions of the integral operator in Eq. (7.1) when $\mathcal{W}\mathcal{T}$ takes the values 0.86, 1.28, 2.06 and $\mathcal{T} = 1$. If $\mathcal{T}$ is not 1, the time axis is scaled by $\mathcal{T}$.

The inverse relationship is often more useful:

$$\gamma = 1 - \left[\cos \left([\cos^{-1}(\sqrt{\lambda_0})]/2 \right) \right]^2 \approx (1 - \lambda_0)/4. \qquad (7.5)$$

By trial and error, we next find the eigenfunction for $\mathcal{W}\mathcal{T}$ and eigenvalue for which Eq. (7.5) gives the required γ value. Then, the optimal base pulse is given by

$$h_{\text{opt}}(t) = \sqrt{\frac{\gamma}{1 - \lambda_0}} \, \psi_0(t) \approx \psi_0(t)/2, \qquad |t| > \mathcal{T}$$

$$= \sqrt{\frac{1 - \gamma}{\lambda_0}} \, \psi_0(t) \approx [1 + 3\gamma/2] \, \psi_0(t), \qquad |t| \le \mathcal{T}. \qquad (7.6)$$

The approximations hold for $\gamma \le 0.02$. A program to find the eigenfunction, pulse and γ for a $\mathcal{W}$ and $\mathcal{T}$ appears in Appendix A.

Example 7.1   *(h_{opt} for 99, 99.9, and 99.999% PIB)*
PSWF eigenfunctions are found by the Appendix program until the γ produced by Eq. (7.5) takes the respective values $0.01, 0.001, 0.00001$. Because γ is monotone decreasing as $\mathcal{W}\mathcal{T}$ grows, this is not difficult, and the eigenfunctions are given in Figure 7.2. Figure 7.3 gives the corresponding optimal $h(t)$ pulses and their spectra, found with the aid of Eq. (7.6). It can be verified that the time side lobes outside $[-1/2, 1/2]$ sum to the required γ (i.e., $\gamma/2$ on each side), and the same for the spectral side lobes outside $|\mathcal{W}| \approx 0.86, 1.28, 2.06$. Changing to a new time interval $[-\mathcal{T}/2, \mathcal{T}/2]$ scales the pulses wider by $\mathcal{T}$ and the spectra narrower by $1/\mathcal{T}$.

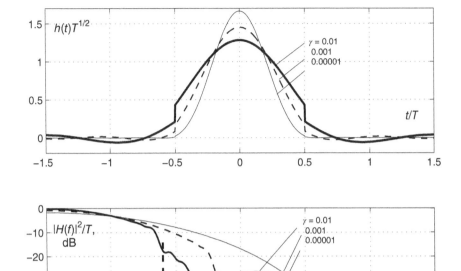

FIGURE 7.3 Optimal standard pulses and their spectra for $\gamma = 0.01, 0.001, 0.00001$ fractional POB in both time and frequency. To convert to a symbol time $T \neq 1$ multiply or divide to remove the T factors shown.

Standard Pulses. Since PSWF pulses are not generally orthogonal for a $\mathcal{T}$, the question arises which one should be associated with the modulator symbol time T. In principle, there is no difference between scaling a $\mathcal{T}$ to fit T and another type of scaling, such as an FTN τ; only the product $\mathcal{WT}$ really matters. The pulses just found in the example are a good fit to a width-1 symbol interval when γ is a reasonable size, and we will take them to be a *standard PSWF modulation pulse* at the $\mathcal{W}$. The figure shows symbol time 1; for time T the standard pulse can be scaled by T as suggested in the figure.

Duality

For a given POB fraction γ, only the corresponding product $\mathcal{WT}$ matters, in the sense that scaling $\mathcal{T}$ scales the PSWF pulse $h(t)$ the same in time and the transform $H(f)$ (and also $\mathcal{W}$) in frequency by the inverse. Furthermore, if $h(t)$ is a PSWF pulse, its transform $H(f)$ is also a time PSWF pulse $H(t)$, whose transform is $h(f)$. The time functions $h(t)$ and $H(t)$ form a dual pair. But duality does not discover new pulses, since the Fourier transform of the pulse for time width $\mathcal{T}$ and frequency width $\mathcal{W}$

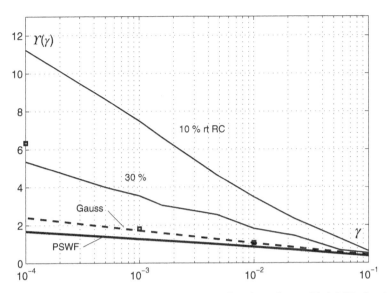

FIGURE 7.4   Time–frequency occupancy Υ as a function of two-sided POB fraction γ. PSWF, Gauss, 10 and 30% root RC isolated pulses. Some IOTA pulse points are shown as squares.

turns out to be the same as the pulse for time $2\mathcal{W}$ and frequency $\mathcal{T}/2$. Both pulses have the same occupancy, namely $\mathcal{W}\mathcal{T}$.

Each time and frequency fraction γ corresponds to its own occupancy. The relationship is plotted in Figure 7.4. The PSWF curve has the least occupancy of any isolated pulse at each POB fraction.

When $\mathcal{W}$ is set to $\mathcal{T}/2$, the dual of the pulse is equal to itself, or alternately, the pulse is its own transform. The time duration then satisfies

$$\mathcal{T} = 2\mathcal{W} = \sqrt{2\Upsilon}. \tag{7.7}$$

This is easily verified with the pulse computation program in Appendix A.

7.1.2   Gauss and Gauss-Like Pulses

Since the the PSWF pulse main lobe is closely related to the Gauss pulse, this is a good place to look at that shape. The unit-energy Gauss pulse is given by

$$h(t) = \frac{1}{\pi^{1/4}\sqrt{a}}\ \exp(-t^2/2a^2), \qquad \text{all } t,\ a > 0. \tag{7.8}$$

The Fourier transform, also unit energy, is

$$H(f) = \pi^{1/4}\sqrt{2a}\ \exp(-2\pi^2 f^2 a^2), \qquad \text{all } f. \tag{7.9}$$

The parameter a is the standard deviation of the pulse; by setting a to ηT, one can be said to have a pulse on symbol interval T with deviation η; by setting $a = \eta\tau T$, the pulse takes on FTN time definition (2.51).

The Gauss pulse has the property that the pulse and its transform have the same shape, namely Gaussian. An interesting example occurs when $a = 1/\sqrt{2\pi} \approx 0.399$, for which $h(t)$ and $H(f)$ are the same function. As with the PSWF pulse, there is a need for a standard Gauss pulse to associate with symbol time $T = 1$, and we will take this one.

Some further calculations relate the POB fraction γ to time and bandwidth intervals for this standard pulse. With reference to Figure 7.1, we can set $\gamma/2$ equal to the integral of $h^2(t)$ over $[\mathcal{T}_\gamma/2, \infty)$. Some calculation shows that this is $Q(\mathcal{T}_\gamma/\sqrt{2}a)$, where $Q(\cdot)$ is the unit Gaussian tail integral defined in Eq. (2.10). Similarly, we set $\gamma/2$ equal the integral of $H^2(f)$ over $(\mathcal{W}_\gamma, \infty)$, and find that this tail is $Q(\mathcal{W}_\gamma\sqrt{8\pi}a)$. The outcome is that the consumed time and frequency for POB fraction γ are

$$\mathcal{T}_\gamma = \sqrt{2}a\,Q^{-1}(\gamma/2)$$
$$\mathcal{W}_\gamma = (1/\sqrt{8\pi}a)\,Q^{-1}(\gamma/2), \tag{7.10}$$

where $Q^{-1}(\cdot)$ is the inverse Q function. When $a = 1/\sqrt{2\pi}$, both $\mathcal{T}_\gamma$ and $2\mathcal{W}_\gamma$ are $(1/\sqrt{\pi})Q^{-1}(\gamma/2)$.

The product of $\mathcal{T}_\gamma$ and $\mathcal{W}_\gamma$ is the occupancy of the Gauss pulse, and for any time scaling, that is, for any variance a^2, it is

$$\Upsilon = \mathcal{T}_\gamma\mathcal{W}_\gamma = (1/2\pi)\,[Q^{-1}(\gamma/2)]^2. \tag{7.11}$$

Figure 7.4 plots Υ as a function of γ.

The IOTA Pulse. The isotropic orthogonal transform algorithm, or IOTA, pulse $\iota(t)$ is a modification of the Gauss pulse that is orthogonal. As with the PSWF pulse, there exists a scaling that equals its own Fourier transform. That version is shown in Figure 7.5; the Gauss pulse that was modified to make $\iota(t)$ appears dotted. Because of the similarity to the Gauss standard pulse, we will take this $\iota(t)$ as the IOTA standard pulse, meaning this $\iota(t)$ is associated with modulation symbol time $T = 1$. It is actually orthogonal, but with respect to intervals of length $\sqrt{2}$.

The IOTA occupancy tracks that of the Gauss pulse at larger POB fractions γ because it closely resembles that pulse. For example, at $\gamma = 0.01$ the IOTA, Gauss, and PSWF occupancies are respectively 1.07, 1.06, and 0.86. Some other values are shown in Figure 7.4. Below $\gamma = 0.005$, the IOTA occupancy is much larger than the Gauss, because of the forced orthogonality.

The IOTA pulse arose in Reference 9 and a rather mathematical tutorial about the subject appears in Reference 10. We can give an overview here. The critical aspect

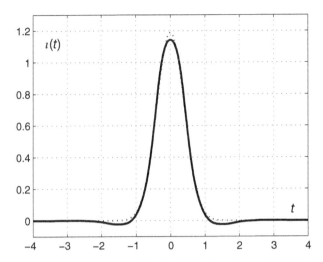

FIGURE 7.5   The IOTA pulse, compared to the originating Gauss pulse (dotted). The Fourier transform is the same, with x-axis f in Hz.

is the *orthogonalization operator*

$$\mathcal{O}x(t) = \frac{x(t)}{\sqrt{\sigma \sum_{k=-\infty}^{\infty} |x(t - k\sigma)|^2}}, \qquad \sigma > 0. \tag{7.12}$$

To start the process, let $x(t)$ be the Gauss pulse in Eq. (7.8) with $a = 1/\sqrt{2\pi}$. The operation $y(t) = \mathcal{O}x(t)$ on $x(t)$ with $\sigma = \sigma_f$ creates a new function $y(t)$ with the property that its transform $Y(f)$ is orthogonal to frequency shifts of $1/\sigma_f$ Hz. However, $y(t)$ is not yet orthogonal to time shifts. To accomplish that, the operator with $\sigma = \sigma_t$ is applied to the transform $Y(f)$, with the outcome that $I(f) = \mathcal{O}Y(f)$ has inverse transform $\iota(t)$ that is orthogonal to time shifts of $1/\sigma_t$ s. Dual orthogonality usually requires that $\sigma_f \sigma_t = 1/2$.[1] It may be helpful to picture the process in operator notation: $\iota(t) = \mathcal{F}^{-1}\mathcal{O}_{\sigma_t}\mathcal{F}\mathcal{O}_{\sigma_f}x(t)$, where $\mathcal{F}$ is the Fourier transform.

When the orthogonality intervals are the same, namely $\sqrt{2}$, we obtain the standard base pulse in Figure 7.5. A unit symbol interval pulse is produced by scaling the pulse narrower by the factor $1/\sqrt{2}$. The IOTA pulse is attracting much interest in future wireless communication because of its advantages in OFDM transmission: It packs pulses relatively well in frequency and time, it has the orthogonality advantage, it resists time and frequency dispersion more equally, and it does not require a cyclic prefix.

---

[1]Under various conditions and start pulses, the procedure can produce orthogonalities of no interest in communication, or it can fail entirely.

Comparison of Occupancies. We are now in a position to compare the occupancy of the PSWF and Gauss pulses to that of the commonly used orthogonal root RC pulses. This is done in Figure 7.4. The Gauss pulse occupancy is surprisingly larger than the PSWF; at 10% POB it is 17% larger, but at POB fraction 0.0001 this grows to 50%, and the IOTA occupancy is four times larger than the PSWF. The reason is that these three pulses are similar only in the main time lobe, and although they differ little in absolute terms in the side lobes, the difference affects the POB at small out of band fractions.

For the familiar 30% root RC orthogonal pulse, occupancy is 42% larger than the optimum at 10% POB and more than three times as much for fractions smaller than 0.0001. This is because a heavy time support penalty must be paid for the pulse's rapid spectral rolloff. The 10% root RC pulse penalty is larger still. This bloated occupancy grows as the pulse excess bandwidth factor drops to zero, and the sinc pulse (factor 0) is unacceptable.

To summarize the situation for an isolated modulation pulse, if power out of band is the criterion, orthogonality exacts a significant occupancy penalty and the sinc pulse is useless. But as the next section shows, when many pulses are optimally combined by linear modulation, pulse time–frequency occupancy is important only with relatively small blocks.

7.1.3   Occupancy of Linear Modulation with FTN

We turn now from the occupancy of an isolated pulse $h(t)$ to the occupancy of an entire modulated signal constructed from $h(t)$. The signal can consist of subcarriers stacked in frequency, and we need to account for this as well as time and frequency scalings of the FTN type. A simple occupancy example with a single subcarrier and 10% root RC modulation was given in Figure 1.2. It shows two time–frequency boxes, an inner one that represents the time and spectral main lobes and an outer one that includes the side-lobe energies (energy inside 99% PIB in the example). The outer box is of interest now; it can be thought of as the time–frequency denied to others.

The analysis that follows takes the "accelerated pulse" view of FTN in Eq. (2.50), not the "stretched pulse" view (these are contrasted in Section 4.1.2). "Accelerated" here means that pulses come faster by the factor τ and do not change otherwise. Their spectrum is unchanged, but subcarriers may be compacted by some factor ϕ. A reference framework based on T-orthogonal pulses without τ and ϕ was given in Chapter 5, Figure 5.1. It was assumed that there is a T-orthogonal base pulse $h(t)$, and the framework was calibrated in steps of T seconds and $1/T$ Hz.

Figure 7.6 now extends Figure 5.1 to FTN and side lobes. It accounts for τ and ϕ, and the pulse spacings are now τT and ϕ/T. If there are N pulses in time and K subcarriers, the inner box has size $N\tau T \times K\phi/T$, an occupancy $\tau\phi NK$ Hz-s. The outer box is wider by $2\varepsilon_t T$ and $2\varepsilon_f/T$, where ε_t and ε_f are respectively the one-sided extra width needed for an isolated $h(t)$ to meet the PIB criterion when $T = 1$. The factor 2 accounts for two sides. When the symbol time is not 1, these extra widths directly scale by T and $1/T$. The $h(t)$ in the ε calculation is a "standard"

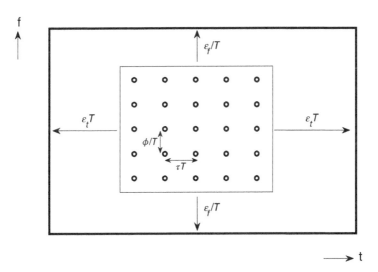

FIGURE 7.6   Reference framework for power out of band (POB) with time and frequency FTN having τ and ϕ squeeze factors.

$h(t)$ with unit symbol time and no FTN scaling. Standard pulses were defined for PSWF, Gauss, and IOTA pulses in Sections 7.1.1 and 7.1.2.

A few assumptions simplify the way forward with little change to the outcome.

   (i) Only peripheral pulses around the inner box contribute to the outer box. This means that the next row inward falls off fast enough that it can be ignored.

  (ii) The base pulse $h(t)$ is symmetric. Its Fourier transform $H(f)$ is thus real and symmetric, and behavior left and right of the inner box as well as above and below is symmetric as well.

 (iii) The same energy fraction γ applies to both the frequency POB and the time POB.

 (iv) The base pulse $h(t)$ can have infinite support (but the part with energy fraction γ is not perceived).

With these simplifying assumptions, it is only the properties of a single standard base pulse that matter. When the symbol time is T instead of 1, the extension to the inner box to the right in Figure 7.6 depends on the standard pulse's right hand time side lobes, and is T times the interval $[1/2, t_o]$ such that the standard pulse's POB after t_o is $\gamma/2$; the extension on the left depends on its left-side lobes and is the same size. The upper extension of the box depends on the standard pulse's positive-frequency side lobes, and is $1/T$ times the frequency interval $[1/2, f_o]$ Hz such that the pulse's POB after f_o is again $\gamma/2$; the lower extension depends on the negative side lobes and is again the same.

The occupancy of the larger box in Figure 7.6 is now

$$(N\tau T + 2\varepsilon_t T)(K\phi/T + 2\varepsilon_f/T) = (N\tau + 2\varepsilon_t)(K\phi + 2\varepsilon_f) \qquad \text{Hz-s.} \quad (7.13)$$

This is independent of T. We can measure occupancy in numbers of symbols instead of seconds and Hz by recalibrating ε_t and ε_f. Let $\varepsilon_t^\tau = \varepsilon_t/\tau$ and $\varepsilon_f^\phi = \varepsilon_f/\phi$. Then the occupancy becomes

$$\phi\tau \, (N + 2\varepsilon_t^\tau)(K + 2\varepsilon_f^\phi) \qquad \text{Hz-s.} \quad (7.14)$$

Note that independent I and Q signals occupy this same time–bandwidth, so that altogether there can be $2NKM$-ary symbols. The occupancy on a per-symbol basis is thus only half the values just given. Consequently, the occupancy created by $h(t)$ modulating at baseband, which counts only positive frequencies, is numerically the same as the occupancy created in an I/Q modulation with the factor 1/2 included. The factor is normally applied in what follows.

For a given bit packet size $\Omega = NK$, the configuration N and K can be *optimized*. This perhaps surprising result stems from Eq. (7.14). If the base pulse has wider side lobes in time, its duration should be longer so that the extra is less per symbol time, and similarly for frequency spread. The outcome will be less occupancy per modulation symbol. Minimization of Eq. (7.14) by simple calculus gives that the minimum occupancy of Ω symbols is

$$\phi\tau \, (1/2)(\sqrt{\Omega} + 2\varepsilon_t^\tau \varepsilon_f^\phi)^2 \qquad \text{Hz-s,} \quad (7.15)$$

where

$$N/K = \varepsilon_t^\tau/\varepsilon_f^\phi \quad \text{or} \quad N = \sqrt{\Omega\varepsilon_t^\tau/\varepsilon_f^\phi} \quad \text{or} \quad K = \sqrt{\Omega\varepsilon_f^\phi/\varepsilon_t^\tau}. \quad (7.16)$$

and the factor of 1/2 accounts for I/Q. For pulses with extreme ε, either N or K is 1. As $\Omega \to \infty$, all ratios N/K lead to the same occupancy per symbol, since only the inner box in Figure 7.6 matters. If there are no subcarriers, the occupancy is simply $(1/2)(N\tau + 2\varepsilon_t)(1 + 2\varepsilon_f)$. Among finite packets and orthogonal pulses, *the sinc pulse does not minimize occupancy*, as will be clear in the next subsection.

Example 7.2 *(Occupancy of 30% Root RC Modulation)*

This commonly used orthogonal pulse has excellent bandwidth properties but a rather long-time extent. This means subcarrier transmission should extend over more time than frequency.

(a) Consider first the no-FTN case. With $\gamma = 0.01$ (99% PIB) and without FTN scaling, the unit symbol time $h(t)$ has $\varepsilon_t = 1.06$ and $\varepsilon_f = 0.0672$ (a total time extent 3.13 s and single-sided frequency width 0.5672 Hz). The ratio $\varepsilon_t/\varepsilon_f$ is 15.8, and this is the optimal ratio K/N of subcarriers to symbol times. For a packet of size $\Omega = 100$, the minimum occupancy is $1 \times 1/2 \times (\sqrt{100} + 2 \times 0.0672 \times 1.06)^2 = 51$ Hz-s, compared to 61 if $K = N = 10$, which is about 20% more. (The factor 1/2 accounts for I/Q). For packet size 10000, these numbers are only 2% apart. With $\gamma = 0.001$ (99.9% PIB), $\varepsilon_f = 0.112$, and $\varepsilon_t = 2.45$, and the savings grow. The optimal ratio K/N is 21.9, and packet size 100 has minimal occupancy 56 compared to 76. With packet size 10000, the minimal occupancy is about 5050, about 3% less than the $K = N$ occupancy.

(b) Next consider the case then $\tau = \phi = 0.7$. With FTN applied, differences are more dramatic. The N/K ratio is the same, but both ε_t^τ and ε_f^τ grow by $1/0.7$. For packet size 100 and 99.9% PIB, the minimum and the $K = N$ occupancies are $0.7^2 \times 1/2 \times (\sqrt{100} + 2 \times 0.1119 \times 2.45/0.7^2)^2 = 32$ and 43 Hz-s. With packet size 10000, these numbers are 2500 and 2630 Hz-s.

7.1.4 PSWF and Gauss Linear Modulation with FTN

The attractive pulses derived earlier can be accelerated in time and squeezed in frequency just as were the more traditional orthogonal pulses. The subject is relatively undeveloped at this writing but Mazo limits have been computed and a few practical applications have been reported.

Interesting areas remain to be explored. Some thoughts and predictions are as follows:

(i) In thinking about receivers, it is good to keep in mind that FTN distance behavior is set mostly by the pulse main lobe, while the side lobes primarily affect the signal occupancy. The main lobes of root RC, PSWF, and Gauss pulses are all similar, and their FTN behavior will not vary much.

(ii) Gauss and PSWF pulses have a better balance of time and frequency; consequently, they are narrower in time but wider in frequency than RC pulses. The optimal simple basis (OSB) receiver structure, which works so well for root RC pulses, requires that symbol-time sampling reconstruct the pulse with reasonable accuracy (Section 2.2.1). A PSWF pulse with POB fraction γ equal, say, 0.01 can be taken to have a practical bandwidth of about $0.86/T$ Hz, considerably more than the nominal bandwidth $0.5/T$ of a root RC pulse. Thus, the sampling interval in time FTN, that is, τT, needs to be about $T/2$ or less with an OSB receiver. If this is inconvenient, new receiver designs can be explored, especially fractional sampling methods.

(iii) The Gauss and PSWF pulse are mid phase, but FTN reception, with its strong ISI, prefers minimum phase (Section 4.1.2). Minimum phase versions can be obtained from these pulses by the MATLAB program recps or another method, and then transmitted. By definition, all phase versions have the same spectrum (Section 2.4), but their duration in the Slepian sense is somewhat extended. A better alternative is to transmit the usual mid-phase pulses in the channel and reduce them to minimum phase before the receiver by an allpass filter (actually, extend them to maximum phase and reverse them).

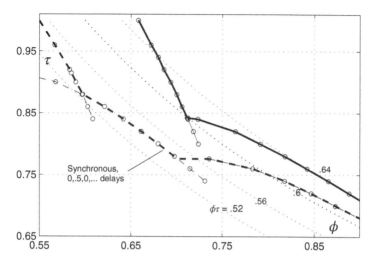

FIGURE 7.7   Estimated Mazo limit positions in the ϕ–τ plane for the standard Gauss pulse. Solid limit is for general case, dashed limit is for alternate subcarriers delayed 0.5 symbol. Contours of constant $\phi\tau$ product are dotted. (Data taken from Reference 11 and private sources.)

Mazo Limit for Gauss Pulses. For binary modulation symbols and time-only FTN acceleration, the Mazo limit for the standard Gauss base pulse, Eq. (7.8) with $a = 1/\sqrt{2\pi}$, is $\tau = 0.665$.[2] That is, $d_{min}^2 = 2$ for the signals generated by $\sqrt{\tau}h(\tau t)$, for all $\tau \geq 0.665$. The limit is the same with 4-ary modulation. The critical error event difference in both cases is [2 -2]. The IOTA pulse, which has a very similar main lobe, has limit 0.653 in both cases and the same event difference.

The occupancy Υ of the base pulse alone for POB fraction $\gamma = 0.01$ is 1.06, no matter what the time scaling. With the standard base pulse ($T = 1$), Υ is divided into a pulse time and frequency widths

$$\mathcal{T} = (1/\sqrt{\pi})\,Q^{-1}(0.01/2) = 1.453 \qquad \text{seconds}$$
$$\mathcal{W} = \mathcal{T}/2 = 0.727 \qquad \text{Hz}$$

for a product 1.06. Scaling the pulse twice as wide, for symbol time 2 or for FTN reasons, gives 2.90 s and 0.364 Hz, with the same product.

For time–frequency FTN with subcarriers, the two-dimensional Mazo limit in the τ–ϕ plane has been computed [11] and some of the data appears in Figure 7.7 (the calculation technique is in Section 5.2). The figure can be compared to Figure 5.7, which is for root RC pulses. As with those pulses, the limit improves if attention is paid

---

[2]This can be computed easily with program `mlsedist2` in Appendix 2A. Note that the calculation must work with samples of the continuous pulse; a sampling rate `fs` equal 10 is sufficient. The calculation first appeared in [8].

TABLE 7.1 Best Known FTN Products on the Mazo Limit Curve for Four Standard Base Pulses

$h(t)$	ϕ	τ	$\phi\tau$
Gauss	0.706	0.86	0.607
PSWF	1.15	0.52	0.598
30% rtRC	0.674	0.89	0.600
10% rtRC	0.660	0.84	0.556

to phase in the subcarriers. The dashed-line limit applies when alternate subcarriers are delayed a half symbol time; its least $\phi\tau$ product, and hence least occupancy, improves by 10%. Overall, the Mazo limits for Gauss pulses are similar to root RC limits, although the best Gauss limits are about 5% improved. The different parts of the limit curves stem from different event families (Section 5.2.2), and some of the component curves are extended to show this more clearly.

Comparison to Root RC Pulses. Table 7.1 compares the binary optimal time–frequency Mazo limit configuration for root RC, Gauss, and PSWF pulses, that is, the point on the limit with least product $\tau\phi$. The values there are based on searches by the author over difference error events at many combinations of τ and ϕ. The modulation base functions $h(t)$ compared are 10 and 30% root RC, Gauss with $\sigma = 0.399$ and the PSWF for POB fraction $\gamma = 0.001$. The standard $h(t)$ are employed, and since custom scales these rather differently, there is a wide variation in the optimal τ and ϕ. However, the product $\tau\phi$ is independent of the particular scaling, and it interesting to observe that the optimal product—the smallest product for which all difference events apparently lead to $d^2 \geq 2$—is quite similar and near 0.6. This leads to the transmission density $2/0.6 = 3.33$ bits/Hz-s. The reason for the similar behavior is likely that the pulse main lobes are similar, and these dominate the Mazo limit outcome.

7.2 SAID'S OPTIMUM DISTANCE PULSES

Another line of analysis based on power out of band seeks to maximize the minimum distance $d_{\min}$ of the modulator's ISI. This distance and its relation to ISI were introduced in Section 2.5, and the central result there is that the symbol error rate of an optimal demodulator is $\approx Q(\sqrt{d_{\min}^2 E_b/N_0})$, with E_b here the energy per bit of the modulator symbols. The problem in this section can be stated: For M-ary modulation pulse $h(t)$ and POB fraction γ, what $h(t)$ maximizes $d_{\min}$?

This question was primarily studied by A. Said [12,13]. His key discovery was that the solution can be found by linear programming. The array of techniques in that field lead quickly to answers. The outcome is tap sequences of the form $\{c_\ell\}$ introduced for root RC pulses in Sections 2.4 and 2.5 and used throughout the book. But the new pulses have higher $d_{\min}$ for their POB than we have seen so far. Said's method is the subject of Section 7.2.1 and some of his optimal tap sets are discussed in Section 7.2.2.

The new tap sets $\{c_\ell\}$ are designed to be used just as those for root RC pulses were in earlier chapters: The analog modulator base pulse is created from $h(t) = \sum c_\ell \varphi(t - \ell T)$, where T is the modulator symbol interval and $\varphi(t)$ is a suitable T-orthogonal interpolation pulse. Because the taps have bandwidth much less than $1/2T$ Hz, the simple OSB receiver structure in Section 2.2 is suitable, with its filter matched to $\varphi(t)$. Asymptotically in E_b/N_0, the new pulses lead to better demodulation at a given POB. Of this there is no doubt. Their behavior with BCJR soft demodulation and poor E_b/N_0 in the early stages of iterative decoding is another question that needs investigation. Some surprises await: Some of the good minimum distance can reside in the out-of-band power, a phenomenon called escaped minimum distance.

7.2.1   Linear Programming Solution

What follows are the high points of Said's linear programming (LP) solution for best pulse $h(t)$ at a fractional POB γ. The method is widely applicable and more details can be found [14], Chapter 6; full details appear in Said's thesis [12].

The key to the method is that distance and spectrum are both linear functions of the pulse autocorrelation. The method solves for that, then finds a tap set from the zeros of the autocorrelation. Let us focus on the case where $h(t)$ can be reproduced by its symbol–time samples $c_\ell = h(\ell T)$.

(i) That the minimum distance of the tap set is a linear function of the autocorrelation of $\{c_\ell\}$ was introduced in Section 2.4. The normalized Euclidean distance between two signals with modulator symbol differences $\{\Delta u_\ell\}$ is in fact

$$d^2 = (1/2E_b) \int \left| \sum_n \Delta u_n h(t - nT) \right|^2 dt = (1/2E_b) \sum_n \sum_k \Delta u_k \rho_h[n - k] \Delta u_n^*,$$

(7.17)

with $\rho[\cdot]$ the discrete-time autocorrelation.

(ii) The PSD of the modulated signal $s(t)$ is given by

$$|S(f)|^2 = \left| \mathcal{F}\left\{ \sum_\ell c_\ell h(t - \ell T) \right\} \right|^2 = |H(f)|^2 \left| \sum_\ell c_\ell \exp(-j2\pi\ell fT) \right|^2$$

$$= |H(f)|^2 \sum_k \rho[k] \exp(-j2\pi k fT).$$

(7.18)

This again is a linear function of $\rho[\cdot]$.

(iii) Since the fractional POB γ will be the constraint, we wish to express that in terms of $\rho[\cdot]$. By the POB definition,

$$\gamma = \int_{|f|>W} (1/T)|S(f)|^2 df$$

(7.19)

(assume the entire integral is unity). Define the discrete-time working variable

$$\chi[k] = \int_{-W}^{W} |H(f)|^2 \exp(-j2\pi kfT)\,df. \tag{7.20}$$

After some manipulation, the POB constraint becomes

$$1 - \gamma = \sum_k \rho[k]\chi^*[k]. \tag{7.21}$$

This expresses γ as a function of a bandwidth $[0, W]$. It may be necessary to solve the LP several times to find the W that connects to a desired γ.

The following semi-infinite dual-form LP now solves for the largest d_{min} *for the given error difference*:

$$\underline{\text{Find}} \qquad d_{opt}^2 = \max_\rho \quad d^2(\rho, \Delta u), \tag{7.22}$$

in which $d^2(\rho, \Delta u)$ is Eq. (7.17), subject to the linear constraints

$$\sum_k \rho[k]\chi[k]^* = 1 - \gamma \qquad \textit{Bandwidth} \qquad \text{[Eq. (7.21)]}$$

$$\rho[0] = 1 \qquad\qquad \textit{Total energy}$$

$$\sum_k \rho[k]\exp(-j2\pi kfT) \geq 0, \quad f \in [0, 1) \quad \textit{Admissibility} \qquad \text{[Eq. (7.18)]}$$

The third constraint means the PSD is positive real. In the LP solution, the constraint ≥ 0 is replaced by $\geq \epsilon$, ϵ small, to guard against numerical errors. There are actually infinitely many constraints here.

The search for the optimal $\rho[k]$ is not complete because we must optimize over the error sequences Δu that may present themselves. It is usually the case that only a few sequences can affect the ultimate minimum distance, for example, short sequences or zero-sum sequences or a set known from a previous optimization. These may be placed in a set S of suspicious differences and the following strategy adopted. Define a new δ and require that $d^2 = d^2(\rho, \Delta u) \geq \delta$ for a given Δu. Modify the LP to

$$\underline{\text{Find}} \qquad d_{opt}^2 = \max_\rho \min_\delta \quad \delta \tag{7.23}$$

with the new constraints

$$d^2(\rho, \Delta u) \geq \delta, \quad \text{all } \Delta u \in S,$$

in addition to the constraints in Eq. (7.22).

We have now the optimum ρ, provided that the error sequence that actually leads to the minimum distance for ρ is in S. That can be tested by finding the minimum distance for ρ, with no restriction on Δu. Several efficient algorithms exist for this purpose, and a summary of these can be found in References 12 or 14, Section 6.3.2. If a sequence not in S turns out to cause a smaller distance than δ, the new error sequence is added to S and the LP is repeated.

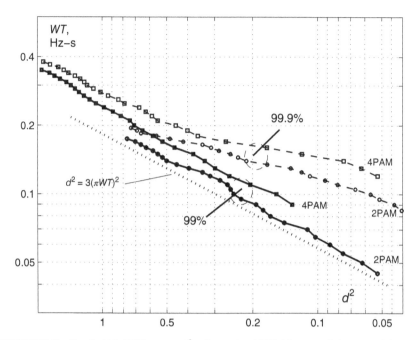

FIGURE 7.8   Bandwidth WT versus $d^2_{\min}$ for optimal PIB binary and 4-ary modulator tap sets found by Said; 99 and 99.9% (dashed) PIB. Quadratic approximation to d^2 is shown dotted. (Data abstracted from Reference 14, Appendix 6A).

In actuality, finding a minimum distance and solving an LP are each an art in themselves. The fine points and some generalizations are omitted here and can be found in References 12 or 14 and elsewhere.

7.2.2   Optimal Modulation Tap Sets

Results of the foregoing LP calculation are shown in Figure 7.8 as WT (Hz-s) versus $d^2_{\min}$, for 15-tap binary and 4-ary modulation with 99 and 99.9% PIB. For comparison, binary sinc pulse modulation has $WT = 0.5$ at $d^2_{\min} = 2$, the upper left corner. Said proved lower and upper bounds to the optimal tap set minimum distance, which all track $(\pi WT)^2$ for small WT. An approximate fit of this relationship to the data is shown in the figure; the accuracy is striking. Results are available for shorter tap sets and for 4QAM modulation [12,14,15]. In the last, the taps are complex valued and cross-dependencies are allowed between I and Q components; the outcome is similar to 4PAM, although the modulator input is now two-channel binary.

Figure 7.9 compares the discrete-time responses of several optimal tap sets at different bandwidths. The spectra of these responses have sharp cutoffs at the respective bandwidths. They function in effect as lowpass filters on the simple modulator output, which can be seen from their successively longer responses. All the tap sets

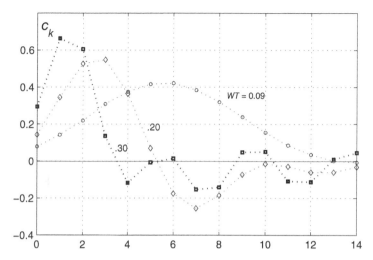

FIGURE 7.9   Optimal modulator tap sets for the 2PAM, 99.9% PIB points at $WT = 0.09, 0.20, 0.30$ in Figure 7.8. Taps are symbol time T spaced.

have much smaller bandwidth than $1/2T$ Hz, so the full modulator base function $h(t)$ can be obtained from these samples with a suitable T-orthogonal interpolation pulse.
    A selection of tap sets is listed in Appendix B.

Escaped Distance.   Because the escaped distance phenomenon often occurs in narrowband signaling, it is worth a side trip to investigate it here. The concept was introduced in Section 2.5.2 and Figure 2.11. By finding the Fourier transform of a difference event signal, we can see how the event's distance is distributed in frequency. It may seem odd that significant distance can exist outside the design bandwidth of a signal set, but signals with narrow bandwidth have high symbol energy and small distance, and a significant part of the distance can fall in the out of band fraction γ even when it is small.

Example 7.3   *(Escaped Distance with a 99% PIB Pulse)*

A rather extreme example occurs with the 15-tap optimal pulse for bandwidth $WT = 0.12$, binary modulation and 99% PIB. The taps $c_\ell, \ell = 0, \ldots, 14$ are given in Appendix B. Using program `mlsedist2` in Appendix 2A, it turns out that $d^2_{\min} = 0.255$ with difference error sequence $\Delta u = [2\ -2\ -2\ 2\ 2\ -2]$ (use `fs = 1` and see Section 2.5.2 for an explanation of error differences). From Eq. (2.53), the power spectrum of the signal error difference is

$$|S_1(f) - S_2(f)|^2 = \left|\mathcal{F}\left\{\sum_k \Delta u[k] h(t - kT)\right\}\right|^2$$

$$= |H(f)|^2 \left|\sum_k \Delta u[k] \exp(-j2\pi k f T)\right|^2, \qquad (7.24)$$

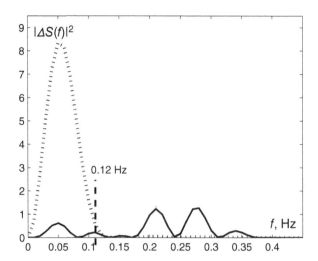

FIGURE 7.10   Power spectral density of the d_{min}-causing error difference event (solid line) for the optimum 15-tap set at $WT = 0.12$ Hz-s and 99% PIB. Full signal PSD is dotted line. Only the positive frequency half of PSDs is shown.

in which $h(t) = \sum_\ell c_\ell \varphi(t - \ell T)$ is constructed from the taps with a suitable T-orthogonal interpolation function $\varphi(t)$. The integral over all f here is d^2_{min} and the integral over $|f| > WT$ is the escaped square distance. Figure 7.10 plots the square distance distribution over frequency, together with the much larger signal PSD $|S(f)|^2$. The square distance that escapes bandwidth $WT = 0.12$ Hz is 0.200, which is 78% of d^2_{min}.

The example shows a dramatic case, because the band limitation is extreme, the POB fraction γ is rather generous and the event multiplicity factor is small. Reducing γ to 0.001 forces more of the minimum distance inside the bandwidth constraint; for $WT = 0.12$ Hz-s, d^2_{min} becomes 0.053, a small number, but most of this lies inside $[-0.12, 0.12]$ Hz-s. It is approximately equal the square distance that does not escape in the 99% case. The effect of escaped distance can be included in the LP optimization and in a search for minimum distance for a tap set by replacing the ordinary distance measure with the spectral one.

Whether escaped distance is a problem depends on what happens to signal components outside the PIB bandwidth constraint. If there is a severe channel filter at that bandwidth, or if there is heavy interference, the signaling will indeed act like its minimum distance lacks the escaped part.

7.2.3   Coded and Uncoded Error Performance

Both earlier chapters and Section 7.2.2 offer methods to reduce modulator spectrum and energy. The goal of this section is to compare Section 7.2.2 signals to the earlier ones. It is challenging to do so because spectrum in this chapter is measured by the POB criterion, whereas earlier spectra utilize the half-power method. One strategy is

TABLE 7.2 d^2_{min} at Nominal 2 and 4 Bits/Hz-s for PAM Modulation Reference, 30% Root RC Pulse with $\tau = 1/2$, and the Optimal-Distance Pulse. POB fraction 0.01 and 0.001. Parentheses Indicate dB Gain Over Reference.

$h(t)$	2 b/Hz-s (0.01)	4 b/Hz-s (0.01)	2 b/Hz-s (0.001)	4 b/Hz-s (0.001)
PAM Ref.	0.8	0.094	0.8	0.094
30% rtRC	1.01 (1.0)	0.154 (2.1)	1.01 (1.0)	0.154 (2.1)
Optimum	1.30 (2.1)	0.503 (7.3)	1.19 (1.7)	0.28 (4.8)

to equate POBs. The 30% root RC pulse measures $0.567\,\tau/T$ and $0.612\,\tau/T$ Hz at $\gamma = 0.01$ and 0.001 fractional out of band power and time-FTN τ, taking the constant symbol view of Eq. (2.51). Given a τ, one can match up an optimal tap set that has the corresponding bandwidth and POB.

Uncoded Transmission. For POB fraction 0.01, a convenient match in parameters occurs when the 30% root RC pulse has FTN $\tau = 1/2$ and the optimal tap set is designed for $0.283/T$ Hz (that is, 0.283 Hz-s). The FTN case has bandwidth $1/2 \times 0.567/T = 0.283/T$ at POB 0.01. Minimum distances can then be compared; they predict the error performance $\approx Q(\sqrt{d^2_{min}E_b/N_0})$ in uncoded transmission, and affect the speed of convergence in the coded case. (How to calulate distance appears in Appendix 2A and Example 2A.3).

As was done in Chapter 4, these modulations can be compared to the simple modulation reference that has the same bits/Hz-s. For the 2- and 4-ary cases, these are 4PAM at 4 bits/Hz-s and 16PAM at 8 bits/Hz-s; coupled to a 30% root RC pulse, they have the same $0.567/T$ Hz bandwidth at POB 0.01.[3] (A full 4PAM reference line can be seen in Figure 4.15) A comparison of all these distances is shown in Table 7.2. Decibel gain over the reference is shown in parentheses; this is the dB reduction in the asymptotic E_b/N_0 required to achieve a symbol error rate (SER).

For POB 0.001, the $\tau = 1/2$ root RC pulse has bandwidth $\approx 0.305/T$ Hz and the optimal tap set is designed for the same. The root RC and reference distances are unchanged. Distances and gains appear in the table.

Table 7.2 shows that moderate but significant gains are available from uncoded optimal tap sets at 2 bits/Hz-s, both over simple modulation and over traditional root RC pulses that are FTN-accelerated. At 4 bits/Hz-s, the gains grow much larger. One can make comparisons at other FTN τ, with roughly similar results. This improvement at higher bit densities is similar to that shown in Figures 4.14 and 4.15.

Figure 7.11 compares measured SER of the 4-ary modulation with base pulse $h(t) = \sqrt{4/9}\,f(4t/9)$, $f(t)$ the 30% root RC pulse, to the SER of the optimal pulse for POB 0.01 and bandwidth $0.250/T$. These two pulses have approximately the

---

[3]The bit densities given here and in the table are kept nominal for easier comparison to earlier chapters. The true densities at POB 0.01 are 4 and $8 \times (0.5/0.567)$, or 3.53 and 7.05 bits/Hz-s; at POB 0.001 they are 3.27 and 6.54 bits/Hz-s.

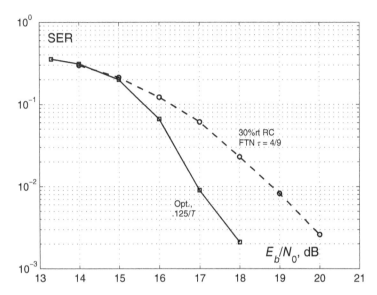

FIGURE 7.11 4-ary modulation SER versus E_b/N_0 in dB for the optimum 15-tap set and the set derived from 30% root RC with FTN $\tau = 4/9$. Both have POB 0.01 bandwidth $W = 0.250/T$ Hz.

same 1% POB bandwidth, and respectively $d^2_{min} = 0.137$ and 0.342. These distances lie nearly 4 dB apart asymptotically and the performance difference is obvious in the figure.

Mazo Limits. A measurement in this spirit can be performed by finding the least bandwidth at which optimal tap sets achieve the matched filter bound distance d_0. For POB 0.01, it is about $0.36/T$ Hz for both 2- and 4-ary modulation (compare to $0.567/T$ for 30% root RC simple modulation). For POB 0.001, it is $0.39/T$ and $0.40/T$ Hz, respectively (compare to $0.612/T$).

Coded Transmission. Recall from Chapter 4 that the progress of iterative decoding breaks into three phases, an initial threshold, which is set by the modulator ISI–convolutional code interaction over a very poor channel, a convergence phase, set mostly by the modulator ISI distance distribution, and a final outcome, set by the convolutional code's error rate on an ISI-free channel with the same E_b/N_0. The optimal-distance modulator $h(t)$ in this section can at most speed the second phase, the convergence. The first stage depends on more subtle properties than minimum distance.

In order to compare coded time-FTN transmission, we match up root RC and optimal distance tap sets by the same procedure as before. A mid-range parameter choice with excellent bandwidth efficiency is 4-ary time FTN with $\tau = 4/9$ applied to a 30% root RC pulse on the one hand, and Said's optimal pulse at $WT = 0.250$ on the other. This is the same comparison as shown in Figure 7.11. Both modulations achieve nominal 6 bits/Hz-s, and have the same 1% POB bandwidth, ≈ 0.250 Hz-s. The root-RC ISI has $d^2_{\min} = 0.137$ and the optimal taps have 0.342, a 4 dB gain. With the same convolutional codes and maps as in Chapter 4, Figure 4.16, the error preformance turns out to be similar to that figure, but about 0.3 dB worse in the threshold region, that is, the needed E_b/N_0 there moves rightward. The reason is that the optimal taps have a poorer overall distance distribution. The improved $d_{\min}$ value leads to a somewhat improved decoder BER when E_b/N_0 is high enough to drive the BER below 10^{-4}.

7.3   CONCLUSIONS

Slepian PSWF base pulses have optimal time–frequency occupancy, which is much lower than competing pulses. Some of this carries over into subcarrier linear modulation, especially if it makes use of FTN techniques.

Gauss and IOTA pulses are similar in their time main lobes and lead to good savings unless the power out of band fraction γ is very small.

Said's optimal $d_{\min}$ pulses can lead to several decibels' improvement in modulator error performance at higher SNR values. The price paid for Gauss, PSWF, and Said pulses is loss of base pulse orthogonality. IOTA pulses retain this.

The escaped distance phenomenon occurs because the difference energy between signals lies outside the nominal design bandwidth. The loss can be a serious factor.

APPENDIX 7A: CALCULATING THE PSWF

Program 7A.1 PSWF and Allied Quantities

Pulses that minimize time–frequency occupancy are related to prolate spheroidal wave functions (PSWFs). This appendix presents a MATLAB function that solves for the PSWF that stems from the time and bandwidth limits $\mathcal{T}$ seconds and $\mathcal{W}$ Hz. In the program, these are called W and L, the latter chosen to avoid confusion with the modulation symbol time T. The program goes on to compute needed parameters, and in particular it calculates the related standard modulation base function $h(t)$.

As developed in Section 7.1.1, a PSWF in our context is the principal eigenfunction of the linear integral operator in Eq. (7.1). It depends only on the product $\mathcal{W}\mathcal{T}$ in the sense that solutions with the same product but $A\mathcal{T}$ are simply scaled in time by A. The function [gam,ev,ef,pulse]=pswfpuls(L,W,t) computes the eigenfunction ef and its eigenvalue ev when $\mathcal{T}$ and $\mathcal{W}$ are L and W and the desired time base for the eigenfunction is t. It also computes the modulation pulse for which the minimal energy fraction gam in time and frequency (γ in Section 7.1.1) lies outside time $[-L/2, L/2]$ and $[-W, W]$. The fraction γ is a consequence of the choice of L and W; if a certain fraction is desired, solutions for successive L and W must be performed until γ is reached. Higher products lead to smaller γ. The computation of γ uses Eq. (7.5).

By setting L to 1, the standard base pulse is obtained, that is, the modulation pulse $h(t)$ that corresponds to symbol time $T = 1$ when there is no FTN time modification. The time base t must be uniform and must include the interval $[-L/2, L/2]$. Larger γ require longer time bases.

The routine works by iterating the operator (7.1) to find the PSWF. The iterations begin with the Gaussian approximation Eq. (7.2), applied to the times in $[-L/2, L/2]$. Because of the nature of the operator's eigenvalues, the convergence slows as the product LW grows, but then the Gaussian start function becomes very accurate. The outcome is that convergence occurs in 1–3 iterations for useful values of γ. An alternate way to find a PSWF is with the MATLAB function dpss, which expresses the problem in discrete time and then finds the eigenfunctions of a certain matrix. Only a part of the modulation pulse can be found this way, but the outcome can be used to initiate the linear operator iterations.

Example calculations appear in Section 7.1.1.

```
function [gam,ev,ef,pulse]=pswfpuls(L,W,t)
%  Function [ev,gam,ef,pulse]=pswfpuls(L,W,t) finds the PSWF 'ef' for time
%  [-L/2,L/2] and bandwidth [-W,W]. It finds also the corresponding unit-energy
%  'pulse', the time & spectral POB fraction gamma 'gam' and the eigenvalue 'ev'.
%  Specify desired uniform time base 't' as a row; it must include [-L/2,L/2]
%  and time 0.

int=t(2)-t(1); pts=length(t);                    %Initialize
ef=zeros(1,pts); pulse=ef;
if t(1)>-L/2 | t(end)<L/2,                        %Illegal timebase?
   disp('ILLEGAL TIMEBASE'), return, end
rg=find(t>=-L/2 & t<=L/2); tr=t(rg); lg=length(rg);   %Times in [-L/2,L/2]
lt=rg(1)-1; rt=rg(end)+1; ctr=floor((lg+1)/2);    %Element locations
if abs(tr(ctr))>1e-6,                             %Check consistency
   disp('INCONSISTENT CENTER TIME'), return, end
disp(['W,L,WL =  ',num2str([W L W*L])])

%                    Initialize recursion with Gaussian
sg2=2*L/(4*pi*W);                                 %2*variance
y=exp(-(tr.^2)/sg2)/sqrt(pi*sg2);                 %Gauss fn.
y=y/sqrt(int*sum(y.^2));                          %Unit energy
%plot(tr,y,'g')                                   %Show Gauss fn.
%              Iterate PSWF operator on full 't' until EV converges.
evold=2; ev=1; nm=0; w2=2*W;
while abs(evold-ev)>.000001, nm=nm+1;             %Recursion loop
   for k=1:pts,
      ef(k)=int*sum(w2*sinc(w2*(t(k)-tr)).*y);   end   %PSWF integral
   evold=ev; ev=sqrt(int*sum(ef.^2));
   disp(['Iter. ',num2str(nm),'  EV = ',num2str(ev)])   %Display EV
   ef=ef/ev; y=ef(rg);                           %Re-normalize
end
gam=1-cos(acos(sqrt(ev))/2)^2;                    %Gamma for EV
disp(['Gamma fraction is ',num2str(gam)])

%              Find pulse from full PSWF and gamma fraction.
%     'gam' is fractional time & frequency POB for this EV and pulse.
%     If it is unacceptable, raise/lower it by lower/raising WL product
%     and run again.
co=sqrt(gam/(1-ev)); ci=sqrt((1-gam)/ev); pulse(1:lt)=co*ef(1:lt);
pulse(rt:end)=co*ef(rt:end); pulse(lt+1:rt-1)=ci*ef(lt+1:rt-1);
pulse=pulse/sqrt(int*sum(pulse.^2));             %Norm to unit energy
```

TABLE 7B.1 Optimal Tap Sets with the PIB Constraint

2PAM, 99% PIB

WT	d²																
$WT = 0.30,$	$d^2 = 1.56$	0.410	0.580	0.366	-0.082	-0.040	-0.049	-0.272	-0.403	-0.040	0.168	-0.054	-0.185	0.009	0.200	0.080
0.26,	1.14	0.352	0.631	0.505	0.132	-0.124	-0.018	0.258	0.145	-0.057	-0.067	-0.012	-0.035	-0.154	-0.243	-0.096
0.20,	0.711	0.270	0.487	0.575	0.444	0.180	-0.138	-0.215	-0.140	-0.011	-0.023	-0.051	-0.129	-0.138	-0.065	0.044
0.15,	0.376	0.208	0.331	0.414	0.393	0.266	0.080	-0.182	-0.353	-0.387	-0.309	-0.139	-0.041	0.077	0.106	0.058
0.12,	0.253	0.157	0.251	0.299	0.339	0.348	0.276	0.147	-0.023	-0.185	-0.299	-0.349	-0.323	-0.276	-0.221	-0.119
0.09,	0.132	0.142	0.105	0.194	0.284	0.290	0.378	0.372	0.372	0.346	0.327	0.242	0.188	0.131	0.085	0.044

2PAM, 99.9% PIB

WT	d²																
$WT = 0.36,$	$d^2 = 1.60$	0.390	0.538	-0.211	-0.596	0.015	0.225	-0.074	0.030	0.019	-0.247	-0.065	0.150	-0.035	-0.110	-0.008
0.30,	1.17	0.294	0.663	0.605	0.136	-0.118	-0.007	0.014	-0.154	-0.142	0.048	0.051	-0.109	-0.113	0.009	0.045
0.26,	0.840	0.218	0.541	0.630	0.303	-0.150	-0.295	-0.109	0.044	-0.034	-0.159	-0.119	0.019	0.073	0.025	-0.007
0.20,	0.454	0.143	0.345	0.527	0.548	0.364	0.070	-0.175	-0.255	-0.186	-0.073	-0.014	-0.029	-0.060	-0.060	-0.032
0.15,	0.118	0.090	0.197	0.312	0.385	0.377	0.269	0.084	-0.124	-0.293	-0.375	-0.362	-0.280	-0.170	-0.072	-0.016
0.12,	0.052	0.079	0.142	0.219	0.308	0.375	0.417	0.421	0.384	0.320	0.238	0.154	0.085	0.035	0.001	-0.007

4PAM, 99% PIB

WT	d²																
$WT = 0.15,$	$d^2 = 0.551$	0.429	0.610	0.368	-0.034	-0.040	0.006	-0.189	-0.241	-0.010	0.025	-0.242	-0.177	0.125	0.319	0.085
0.12,	0.315	0.315	0.446	0.448	0.041	-0.263	-0.316	-0.093	0.012	-0.057	-0.125	0.035	0.289	0.371	0.277	0.055
0.09,	0.193	0.262	0.406	0.509	0.421	0.140	-0.167	-0.331	-0.342	-0.157	0.003	0.093	0.033	-0.032	-0.105	-0.097
0.065,	0.103	0.179	0.209	0.303	0.354	0.387	0.239	0.121	-0.042	-0.199	-0.337	-0.352	-0.323	-0.260	-0.160	-0.095
0.045,	0.053	0.142	0.106	0.195	0.284	0.290	0.380	0.373	0.371	0.347	0.326	0.240	0.187	0.133	0.084	0.041

4PAM, 99.9% PIB

WT	d²																
$WT = 0.15,$	$d^2 = 0.254$	0.269	0.543	0.405	-0.118	-0.336	-0.095	0.022	-0.254	-0.418	-0.182	0.112	0.155	0.109	0.082	0.066
0.12,	0.109	0.181	0.413	0.547	0.387	0.036	-0.267	-0.334	-0.242	-0.167	-0.149	-0.094	0.018	0.137	0.142	0.080
0.09,	0.044	0.122	0.277	0.439	0.529	0.489	0.299	0.078	-0.101	-0.142	-0.068	0.061	0.140	0.163	0.117	0.052
0.065,	0.017	0.081	0.149	0.246	0.346	0.423	0.449	0.430	0.359	0.258	0.153	0.056	-0.005	-0.032	-0.035	-0.031

APPENDIX 7B: OPTIMUM DISTANCE TAP SETS

What appears in Table 7B.1 are length-15 optimal-d_{min} tap sets c_ℓ, $\ell = 0, \ldots, 14$ computed by A. Said. These are for use in 2PAM and 4PAM modulation with a PIB constraint. They provide the best d_{min} sets for constraints 99 and 99.9% power inside a bandwidth $[-WT, WT]$ specified in Hz-s, or alternately in Hz with symbol time $T = 1$ s. The modulation base function is $h(t) = \sum_k c_\ell \varphi(t - \ell T)$, where $\varphi(t)$ is a suitable T-orthogonal interpolation function, that is, one with flat spectrum in the frequency range of interest.

All shown tap sets are minimum phase. Many further sets, including shorter ones and sets designed for 4QAM modulation, are available in References 13 and 15.

REFERENCES

1. *D. Slepian, On bandwidth, *Proc. IEEE*, **64**, pp. 292–300, 1976.

2. D. Slepian and H.O. Pollak, Prolate spheroidal wave functions, Fourier analysis and Uncertainty—I, *Bell Sys. Tech. J.*, **40**, pp. 43–64, 1961.

3. H.J. Landau and H.O. Pollak, Prolate spheroidal wave functions, Fourier analysis and Uncertainty—II, *ibid.*, pp. 65–84.

4. H.J. Landau and H.O. Pollak, Prolate spheroidal wave functions, Fourier analysis and Uncertainty—III: the dimension of the space of essentially time- and band-limited signals, *Bell Sys. Tech. J.*, **41**, pp. 1295–1336, 1962.

5. D. Slepian, Prolate spheroidal wave functions, Fourier analysis and Uncertainty—IV: extensions to many dimensions; generalized prolate spheroidal functions, *Bell Sys. Tech. J.*, **43**, pp. 3009–3057, 1964.

6. D. Slepian, Prolate spheroidal wave functions, Fourier analysis and Uncertainty—V: The discrete case, *Bell Sys. Tech. J.*, **57**, pp. 1371–1430, 1978.

7. D. Slepian, Some comments on Fourier analysis, uncertainty and modeling, *SIAM Rev.*, **25**, pp. 379–393, 1983.

8. J.B. Anderson and F. Rusek, Optimal side lobes under linear and faster-than-Nyquist modulation, *Proc. IEEE Int. Symp. Inf. Theory*, Nice, pp. 2301–2304, 2007.

9. A. Alard, Construction of a multicarrier signal, U.S. Patent 6,278,686, 2001.

10. B. Le Floch, M. Alard, and C. Berrou, Coded orthogonal frequency division multiplex, *Proc. IEEE*, **83**, pp. 982–996, 1995.

11. F. Rusek and J.B. Anderson, Multistream faster than Nyquist signaling, *IEEE Trans. Commun.*, **57**, pp. 1329–1340, 2009.

12. A. Said, Design of optimal signals for bandwidth efficient linear coded modulation, Ph.D. thesis, Dept. Electrical, Computer and Systems Eng., Rensselaer Poly. Inst., Troy, NY, 1994.

13. A. Said and J.B. Anderson, Bandwidth efficient coded modulation with optimized linear partial-response signals, *IEEE Trans. Inf. Theory*, **44**, pp. 701–713, 1998.

---

*References marked with an asterisk are recommended as supplementary reading.

14. *J.B. Anderson and A. Svensson, *Coded Modulation Systems*, Kluwer-Plenum, New York, 2003, Chapter 6.

15. A. Said and J.B. Anderson, Tables of optimal partial-response trellis modulation codes, Communication, Information and Voice Processing Report Series, No. 93-3, Elec., Computer and Systems Eng. Dept., Rensselaer Poly. Inst., Troy, NY 1993 (available from the publisher website or the author).

INDEX

acceleration factor, τ
  definition, 45-46
  receiver, effect on, 138, 142
aliasing, 35–37, 71. *See also* pulse
allpass filter, 40, 86, 116
antipodal event, 43
antipodal signal bound, 43–44
AP (a priori information). *See also*
      instrinsic subtraction
  in ISI-BCJR, 100
  in iterative decoding, 102, 104
AWGN (additive white Gaussian noise),
      18, 19
  colored noise, 35
  scaling, 110–111

backup recursion, 97
backward recursion, 33
Baker's rule, 156
bandwidth. *See also* spectrum
  conversion of criteria, 73, 157–158, 163
  criteria, 10
  3 dB, 10
  excess, 67-68
  general, 2–4, 8
  half power, 10
  PIB, 10–11, 157 (in CPM), 163

POB, 10–11, 164, 178–179, 183
  wide versus narrow band, 13
Barbieri, A., with Fertonani and Colavolpe,
      142
basis. *See* orthogonality
BCJR (Bahl-Cocke-Jelinek-Raviv)
      algorithm. *See* trellis detection
binary entropy function, 64,
      75–76
bit density
  in capacity calculation, 74–76
  in CPM, 157
  definition, 9, 63
  in frequency FTN, 129, 133
  in FTN, 81, 111–113
  PSD, effect of, 73
  in set-partition coding, 150
bit energy, 5, 30

capacity
  AWGN, 59–61
  bandwidth channel, 61–63
  for BER, 64–68, 73–74
  constrained, 59–61
  for CPM, 157–160
  for linear modulation, 68–72
  for PSD, 64–68

Bandwidth Efficient Coding, First Edition. John B. Anderson.
© 2017 by The Institute of Electrical and Electronics Engineers, Inc. Published 2017 by John Wiley & Sons, Inc.

Shannon limits, 110–113, 132–134, 140, 157

Shannon limit calculation, 63, 73–77
of square PSD, 65
for time-frequency FTN, 76–77, 132–134

CC characteristic, 84, 103–109

CC lines, 111–113

channel model. *See also* phase; receiver
AWGN, 18–19
for FTN, 82, 86–88, 115–116
minimum distance of, 37–43
precursor, 86, 115–116
tail, 86

channel shortening, 93

channel use, 59
in frequency FTN, 132

coded modulation. *See* CPM; set-partition coding

coding. *See also* convolutional codes; CPM; set-partition codes
definition, 13–14
general, 3–4, 13–15
LDPC, 142–143
modulation, compared to, 13–14
parity-check, 13–14
semisystematic, 107

colored noise receiver. *See* Receiver

constant envelope, 146, 157

convolutional codes
CC characteristics, 84, 103–109
CC line, 111–113
in CPM, 158–160
in FTN (general), 82–85
in FTN (good codes), 101–110, 117–122
general, 30, 84–85
generators, definition of, 117–120
semisystematic, 107
in set-partition coding, 151–152

convolution, in CPM, 154

correlation factor, ρ, convolutional code, 103–109, 117–120

correlator receiver. *See* receiver

CPFSK (continuous-phase frequency-shift keying). *See* modulation

CPM (continuous phase modulation) signaling
coded, 153, 158–160
distance, 156
error performance, 158–160
general, 153–160
rate of, 157

Shannon limit, 157–160
spectrum, 155–156

Darwazeh, I., 140

Dasalukunte, D., 140

decision depth. *See* trellis detection

decision region. *See* distances

delay, of pulse. *See* frequency FTN

difference sequence. *See* distance

distance. *See also* Mazo limit
antipodal, 43–44
calculation, 38, 50–54
critical, 49
decision regions, 21
difference sequence, 38, 134
equivalent, in BCJR, 99–100
error events, 37–38, 43–47, 134
escaped, 48–50, 181–182
free, 116–122
in frequency FTN, 134–136
FTN model for, 115–116
minimum, of channel, 37–43
minimum, of 2 signals, 21–23
of modulations, 21–22, 50–54
normalized, 20
optimal distance pulses, 180–181

Douillard, C., 83

EER (event error rate). *See* error rates

equalizers, 31, 80, 142

equivalent distance, in BCJR. *See* distance

error events. *See* distance

error family, 137

error probability. *See also* distance
calculation of, 54–55, 90
Gaussian, 20
of modulations, 21–24

error rates
BCJR, 84–85, 100–101, 105
of bits (BER), 54
of events (EER), 43, 90
measurement of, 54–55
of symbols (SER), 47

error spectra, 48–50, 181–182

escaped distance. *See* distance

excess bandwidth. *See* bandwidth

excess phase, in CPM, 153–154

extrinsic information, 84

Falconer, D.D., with Magee, 93
FTN (faster then Nyquist) signaling. *See
    also* frequency FTN
  capacity for, 81
  channel model, 82, 86, 115–116
  classical FTN, 80–82
  coded, 82–85, 88, 101–110, 117–122
  definitions, 45–46, 81
  distance calculation, 52
  with LDPC, 142–143
  occupancy, 172–175
  receiver for, 82
  spectrum, 81–82
  with Said pulses, 183–185
Fettweis, G., 142
folded spectrum. *See* spectrum
forward recursion, 33
fractional sampling, 30
frequency FTN. *See also* FTN
  definition, 128–133
  delay, of pulse, 135
  distance, 134–136
  Mazo limit, 136–138
  occupancy, 172–175
  receivers, 138–142
  Shannon limit, 131–134, 140
  subcarriers, 128
  synchronous, 135, 138
  time-frequency reference, 128–129,
    172–173
frequency squeeze factor ϕ
  definition, 128–129
  receivers, effect on, 138, 142

Gauss pulse, 142, 169–170
  Mazo limit, 176–177
  occupancy, 169–170, 172, 175
generator, 27, 32, 80
  convolutional code definition, 117–120
GFDM (generalized frequency-division
    multiplexing), 142
Gibby-Smith condition, 36, 71
Gram-Schmidt procedure, 19, 29–30
  receiver, 141
Gray map, 108, 120–123

half power frequency, 10. *See also*
    bandwidth
Hirt, W., with Massey, 68–69
Hz-s (hertz-second), 9

in-phase signal, 6, 129–133
intrinsic subtraction, 96–98, 103, 110

IOTA (isotropic orthogonal transform
    algorithm) pulse, 140,
    170–172
irrelevance, theorem, 19
ISI characteristic, 85, 99–101
iterative decoding, 82–86, 105, 110–113,
    138

Kotelnikov, V.A., 18

LDPC (low density parity-check codes).
    See coding
Lender, A., with Kretzmer, 80
Lindell, G., 159–160
linear modulation, 5
  capacity, defined, 68–72
linear programming. *See* Said pulse
LLR (log-likelihood ratio), 34, 98,
    102

M-algorithm, 80, 93–94
map, 3:2, 122
MAP (maximum a posteriori) receiver. *See*
    receiver
master constellation, 147–148
matched filter bound. *See* antipodal signal
    bound
max-log-MAP algorithm, 34–35
Mazo, J., 45, 80
Mazo limit, 45 (definition), 80, 95
  in frequency FTN, 136, 138
  Gauss pulse, 176–177
  Said pulse, 184
minimum distance. *See* distance
ML (maximum likelihood) receiver. *See*
    receiver
modulation. *See also* distance; linear
    modulation
  baseband
  coding, compared to, 13
  CPFSK, 154–156
  definition, 3–5
  high energy, 13
  index, 154
  ML detection of, 24
  MSK, 156, 159
  non-orthogonal, 25
  octal, 5, 22, 61
  2PAM, 5, 21, 38, 61
  4PAM, 5, 22, 61
  passband, 6
  QAM, 146–147
  QPSK, 23, 61

modulation. *See also* distance; linear
    modulation (*Continued*)
  simple, 21, 24, 61, 71
  wide versus narrow band, 13
modulation index, in CPM, 154
MSK (minimum-shift keying). *See*
    modulation
multicarrier FTN. *See* frequency FTN
multiplicity, of error events, 47
mutual information, 59

Nyquist pulse
  criterion (NPC), 6
  orthogonality, 7, 81

occupancy, Υ
  of FTN, 175–177
  of pulse, 164–172
  of signal, 172–175
OFDM (orthogonal frequency-division
    multiplex), 128, 131
offset, label, 89–92
optimality principle, 30
orthogonal pulse, 7–9
orthogonality
  basis, orthognal, 18, 26–27
  capacity, effect on, 71–72
  in frequency FTN, 129
  in modulation, 21–27
orthogonalization operator, 171
OSB condition, 26 (definition), 37
OSB (orthogonal simple basis) receiver.
    See receiver

PAM (pulse amplitude modulation).
    See Modulation
partial energy function, 40
phase
  minimum etc., definition, 40–41, 175
  super minimum, 115
  versions, 38, 52
PIB (power in band), 10 (definition). *See
    also* bandwidth
POB (power out of band), 10 (definition).
    See also bandwidth
Prlja, A., 93
PRS (partial response signaling), 27, 80
  in CPM, 154–155
PSD (average power spectral density),
    11–12 (definition). *See also*
    Spectrum
PSWF (prolate spheroidal wave function)
  definition, 165–166

occupancy, 164–175, 177
  as pulse, 168–169
  solution for, 166–168 (examples),
    185–186 (program)
pulse. *See also under type Gauss, RC, etc.*
  aliased, 35–37, 71
  duality, 168–169
  standard pulse, 168, 173
QAM (quadrature amplitude modulation).
    See Modulation
Q function (definition), 21
QPSK (quadrature phase-shift keying). *See*
    Modulation
quadrature signal, 6, 129–133

rate
  of code, 14
  in CPM, 157
  in frequency FTN, 132–133
  in FTN, 105, 107, 110
  in set-partition coding, 148
receiver. *See also* iterative decoding; SIC;
    trellis detection
  chips, 140–141
  OSB, 26 (definition), 37, 41–43, 53, 82
  correlator, 27–28
  fractional sampling, 30
  Gram-Schmidt, 29–30
  MAP, 18, 20, 84
  matched filter, 28, 36–37, 41–43
  ML, 18, 20, 99–100
  ML, for modulation, 24–27
  OSB, 26 (definition), 37, 41–43,
    53, 82
  vector, 20
reduced search. *See* trellis detection
reduced trellis. *See* trellis detection
residual, in M-BCJR, 96–97
Robertson, P., 35
root RC (raised cosine) pulse
  definition, 7–9
  occupancy, 169, 174–175, 177
  Shannon limit with, 66–68
  spectrum, 8, 183
Rusek, F., 93, 139

Said, A., 38, 81, 177
Said pulse
  error performance, 183–185
  linear program solution, 178–180
  optimal pulses, 180–181, 187–188
sampling theorem, 25–27
scale factor, AWGN, 110–111

SEFDM (spectrally efficient
    frequency-division multiplex),
    140–141
semisystematic code. *See* convolutional
    codes
SER (symbol error rate). *See* error rates
Seshadri, N., 48, 80
Set-partition coding
    definition, 146–147
    distance in, 147–149
    error performance, 151–153
    Shannon limit, 150–151, 153
    spectrum and rate, 150
Shannon, C.E., 2, 14, 18, 60
SIC (successive interference cancelation),
    31, 138–142
signal space, 18–20
sinc pulse, 7 (definition), 12, 61, 80,
    174
Slepian, D., 61, 165
Slepian's problem, 131, 164–170
slope, convolutional code, 102–109
    (analysis), 117–122
soft information, 84
spectral antisymmetry, 6, 65, 72
spectral factorization, 41
spectrum. *See also* bandwidth
    in CPM, 155–157
    definition, 11
    error event, 48–50, 182
    feature, of PSD, 73
    folded, 35–36, 69
    for FTN, 82, 87–88
    for linear modulation, 12
    PSD, 11–12
    of PSWF, 168
    for set-partition coding, 150
    of Said pulses, 183,
        178–179
sphere decoder, 31, 141–142
subset selector, 147-149
super minimum phase, 115

symbol energy, 6 (definition)
synchronous FTN. *See* frequency FTN

TCM (trellis-coded modulation). *See*
    set-partition coding
Theorem, 2WT, 14, 61–62, 67–68, 128
threshold, iterative, 84
trellis detection. *See also* M-algorithm;
    receiver
    BCJR, 31–35, 91–92
    branch labels, 89–92, 117–122
    for continuous signals, 31
    decision depth, 30, 87, 97
    definition, 29–30
    M-BCJR, 94–98, 111–113
    max-log-MAP, 35
    one-way BCJR, 96
    reduced search, 30, 93–98, 156
    reduced trellis, 30, 89–93
    residual BCJR, 97
    sphere decoder, 31, 141–142
    tail offset BCJR, 91–92
    time offset BCJR, 92–93
tunnel, iterative, 85, 102, 105–110
turbo equalization, 82

Ungerboeck, G., 146
Ungerboeck rules, 148
unit circle, 38–40, 52, 86

VA (Viterbi algorithm), 30, 89–91.
    See also trellis detection
    in CPM, 156
    in set-partition coding, 150
Viterbi, A.J., 30

whitened matched filter, 28–29, 36–37,
    41–43

Zero sum, 49
Zhang, W., with Schlegel, 152–153

IEEE PRESS SERIES ON DIGITAL AND MOBILE COMMUNICATION

John B. Anderson, *Series Editor*
University of Lund

---

1. *Wireless Video Communications: Second to Third Generation and Beyond*
   Lajos Hanzo, Peter J. Cherriman, and Jurgen Streit
2. *Wireless Communications in the 21st Century*
   Mansoor Sharif, Shigeaki Ogose, and Takeshi Hattori
3. *Introduction to WLLs: Application and Deployment for Fixed and Broadband Services*
   Raj Pandya
4. *Trellis and Turbo Coding*
   Christian B. Schlegel and Lance C. Perez
5. *Theory of Code Division Multiple Access Communication*
   Kamil Sh. Zigangirov
6. *Digital Transmission Engineering,* Second Edition
   John B. Anderson
7. *Wireless Broadband: Conflict and Convergence*
   Vern Fotheringham and Shamla Chetan
8. *Wireless LAN Radios: System Definition to Transistor Design*
   Arya Behzad
9. *Millimeter Wave Communication Systems*
   Kao-Cheng Huang and Zhaocheng Wang
10. *Channel Equalization for Wireless Communications: From Concepts to Detailed Mathematics*
    Gregory E. Bottomley
11. *Handbook of Position Location: Theory, Practice, and Advances*
    Edited by Seyed (Reza) Zekavat and R. Michael Buehrer
12. *Digital Filters: Principle and Applications with MATLAB*
    Fred J. Taylor
13. *Resource Allocation in Uplink OFDMA Wireless Systems: Optimal Solutions and Practical Implementations*
    Elias E. Yaacoub and Zaher Dawy
14. *Non-Gaussian Statistical Communication Theory*
    David Middleton
15. *Frequency Stabilization: Introduction and Applications*
    Venceslav F. Kroupa
16. *Mobile Ad Hoc Networking: Cutting Edge Directions,* Second Edition
    Stefano Basagni, Marco Conti, Silvia Giordano, and Ivan Stojmenovic
17. *Techniques for Surviving the Mobile Data Explosion*
    Dinesh Chandra Verma and Paridhi Verma

18. *Cellular Communications: A Comprehensive and Practical Guide*
    Nishith D. Tripathi and Jeffrey H. Reed
19. *Fundamentals of Convolutional Coding,* Second Edition
    Rolf Johannesson and Kamil Sh. Zigangirov
20. *Trellis and Turbo Coding,* Second Edition
    Christian B. Schlegel and Lance C. Perez
21. *Bandwidth Efficient Coding*
    John B. Anderson